ROYALTIES AT RISK IN THE DEVELOPMENT OF THE NATION'S NATURAL RESOURCES

Royalties at Risk in the Development of the Nation's Natural Resources

Thomas Patrick Marinelli
Editor

Nova Science Publishers, Inc.
New York

For permission to use material from this book please contact us:
Telephone 631-231-7269; Fax 631-231-8175
Web Site: http://www.novapublishers.com

LIBRARY OF CONGRESS CATALOGING-IN-PUBLICATION DATA

Marinelli, Thomas.
Royalities at risk in the development of the nation's natural resources / [Thomas Marinelli].
p. cm.
Includes index.
ISBN 978-1-60692-837-0 (hbk.)
1. Oil and gas leases--United States. 2. Mining leases--United States. 3. Natural resources--United States. 4. Energy development--United States. I. Title.
HD9566.M375 2009
333.1'1--dc22
2009035589

Published by Nova Science Publishers, Inc. ✦ New York

CONTENTS

PREFACE

In addition to having a low government take, the deep water Gulf of Mexico and other U.S. regions are attractive targets for investment because they have large remaining oil and gas reserves and the U.S. is generally a good place to do business compared to many other countries with comparable oil and gas resources. Multiple studies completed as early as 1994 and as recently as June 2007 indicate that the U.S. government take in the Gulf of Mexico is lower than that of most other fiscal systems. For example, data GAO evaluated from a June 2007 industry consulting firm report indicated that the government take in the deep water U.S. Gulf of Mexico ranked 93rd lowest of 104 oil and gas fiscal systems evaluated. Generally, other measures indicate that the United States is an attractive target for oil and gas investment. The lack of price flexibility in royalty rates—automatic adjustment of these rates to changes in oil and gas prices or other market conditions—and the inability to change fiscal terms on existing leases have put pressure on Interior and the Congress to change royalty rates in the past on an ad hoc basis with consequences that could amount to billions of dollars of foregone revenue. For example, royalty relief granted on leases issued in the deep water areas of the Gulf of Mexico between 1996 and 2000—a period when oil and gas prices and industry profits were much lower than they are today—could cost the federal government between $21 billion and $53 billion, depending on the outcome of ongoing litigation challenging the authority of Interior to place price thresholds that would remove the royalty relief offered on certain leases. Further, royalty rate increases in 2007 are expected to generate modest increases in federal revenues from future leases offered in the Gulf of Mexico. However, in choosing to increase royalty rates, Interior did not evaluate the entire oil and gas fiscal system to determine whether or not these increases strike the proper balance between the attractiveness of federal leases for investment and appropriate returns to the federal government for oil and gas resources.

Interior does not routinely evaluate the federal oil and gas fiscal system, monitor what other governments or resource owners are receiving for their energy resources, or evaluate and compare the attractiveness of federal lands and waters for oil and gas investment with that of other oil and gas regions. As a result, Interior cannot assess whether or not there is a proper balance between the attractiveness of federal leases for investment and appropriate returns to the federal government for oil and gas resources. Specifically, Interior does not have procedures in place for evaluating the ranking of (1) the federal oil and gas fiscal system or (2) industry rates of return on federal leases against other resource owners. Interior also does not have the authority to alter tax components of the oil and gas fiscal system. All these factors are essential to inform decisions about whether or how to alter the federal oil and gas fiscal system in response to changing market conditions.

Chapter 1 - In fiscal year 2007, domestic and foreign companies received over $75 billion from the sale of oil and gas produced from federal lands and waters, according to the Department of the Interior (Interior), and these companies paid the federal government about $9 billion in royalties for this oil and gas development. The government also collects other revenues in rents, taxes, and other fees, and the sum of all revenues received is referred to as the "government take." The terms and conditions under which the government collects these revenues are referred to as the "oil and gas fiscal system." This report (1) evaluates government take and the attractiveness for investors of the federal oil and gas fiscal system, (2) evaluates how the absence of flexibility in this system has led to large foregone revenues from oil and gas production on federal lands and waters, and (3) assesses what Interior has done to monitor the performance and appropriateness of the federal oil and gas fiscal system. To address these issues, the authors reviewed expert studies and interviewed government and industry officials.

Chapter 2 - The Department of the Interior's (Interior) Minerals Management Service (MMS) collected the equivalent of over $9 billion in oil and gas royalties in fiscal year 2007, more than $5 billion of which it deposited in the U.S. Treasury; it dispersed the remaining approximately $4 billion to other federal, state, and tribal accounts. These royalties—payments made to the federal government for the right to produce oil and gas from federal lands and waters—represent one of the country's largest nontax sources of revenue. The amount of oil and gas royalties MMS collects may increase if the price of energy increases and industry's demand to drill on lands and in waters controlled by the federal government continues to trend upward. For example, the price of West Texas Intermediate—a type of oil commonly used as a benchmark—has risen by more than 100 percent since January 2007 and recently exceeded $140 per barrel, a price that, when adjusted for inflation, is at the highest level since 1980. Moreover, applications for onshore drilling permits at Interior's Bureau of Land Management (BLM) continue to increase, from approximately 4,000 in 2001 to more than 10,000 in 2007. Similarly, offshore leasing activity overseen by MMS's Offshore Energy and Minerals Management (OEMM) has also generally increased since 2001.

Chapter 3 - In recent years, many of us have witnessed a sorry pattern across the spectrum of Federal agencies. At every henhouse a fox is stationed. The well-connected few are rewarded at the expense of the common folk, and all the while the people's branch has been looking the other way.

The Minerals Management Service exemplifies this troubling pattern. Throughout the last several years, the MMS seems to have drifted further and further into the grasp of the oil and gas industry. No doubt this occurred in part because of the agency's dual and conflicting roles.

On the one hand the MMS is charged with developing energy resources on our public lands. On the other hand, it is supposed to collect at a fair value payment owed to people for the development of the resources they own. But something has gone wrong. Development has been thriving. Oil and gas drilling has dramatically increased. It is obvious by the sheer number of rigs that are sprouting up all over the West and drilling platforms in the Gulf of Mexico.

Chapter 4 - Under current law, the hardrock mineral industry pays no royalty to the federal government for the privilege of extracting resources from federal lands. This differs from the federal policy toward the coal and oil/gas industries, the policy of State governments, and the leasing arrangements in the private sector, which often require bonus bids and an ad-valorem royalty on the value of the resulting output. Hardrock mining on acquired federal lands pays a 5% royalty.

The current federal policy toward hardrock minerals is inconsistent with the fundamental market principle that a royalty is a factor payment, part of the rent paid, or the return, to land as both a marketable capital asset and input to production. In general, the free development of federal mining land will result in more public land developed and more minerals produced than is economically efficient. Another implication is that any economic rents, i.e., excess profits to a resource owner above the level required to produce or supply the resource in the long run, would accrue to private rather than public beneficiaries. The free development regime for hardrock minerals on federal lands was created to stimulate economic development of the west and has more recently been sustained to protect the viability of the United States hardrock mineral industry and to prevent negative economic impacts on western communities built around that industry. Introducing a royalty payment system might have an adverse economic effect on hardrock mineral producers but it would also tend to increase output in the rest of the economy and promote a more efficient use of national resources. The hardrock mining industry generally has, over the last four years, been booming, and any adverse industry effects would tend to be mitigated.

The appropriate royalty system, according to economic principles, is the advalorem royalty based on the market value of the mineral upon extraction, adjusted for any externality-related taxes. Using "gross income for depletion purposes" is conceptually the same as market value, although it has practical advantages since that is the basis for a producer's computation of its percentage depletion allowance for tax purposes. Using net smelter return — the basis for most private royalty contracts — is conceptually the same as gross income, although there may be differences due to deductions for costs, and would require a new administrative apparatus. Using the producer's net profits as the basis would render the royalty an income tax, which would be inconsistent with mainstream economic principles that a royalty is part of a factor payment, and would likely reduce, and at times totally eliminate, royalty payments. The U.S. hardrock minerals industry pays income taxes, including the alternative minimum tax, and, in addition, is assessed a variety of claims and patent fees. In addition, hardrock mining firms qualify for some special tax benefits or subsidies: expensing (i.e., a current deduction) of exploration and development costs; the percentage depletion allowance, at rates ranging from 14% to 22% of gross income; and a deduction for mine closing and reclamation costs. The special mining tax breaks are sufficient to lower the effective marginal tax rate slightly below that for other industries. These tax rates, however, are currently much higher than the historical rates, which were either close to zero or negative.

In: Royalties at Risk in the Development...
Editor: Thomas Marinelli

ISBN: 978-1-60692-837-0

Chapter 1

Oil and Gas Royalties: The Federal System for Collecting Oil and Gas Revenues Needs Comprehensive Reassessment *

U. S. Government Accountability Office

Why GAO Did This Study

In fiscal year 2007, domestic and foreign companies received over $75 billion from the sale of oil and gas produced from federal lands and waters, according to the Department of the Interior (Interior), and these companies paid the federal government about $9 billion in royalties for this oil and gas development. The government also collects other revenues in rents, taxes, and other fees, and the sum of all revenues received is referred to as the "government take." The terms and conditions under which the government collects these revenues are referred to as the "oil and gas fiscal system." This report (1) evaluates government take and the attractiveness for investors of the federal oil and gas fiscal system, (2) evaluates how the absence of flexibility in this system has led to large foregone revenues from oil and gas production on federal lands and waters, and (3) assesses what Interior has done to monitor the performance and appropriateness of the federal oil and gas fiscal system. To address these issues, we reviewed expert studies and interviewed government and industry officials.

What GAO Recommends

Interior did not fully agree with the recommendations in our draft report, stating that an ongoing Interior study covered many of the issues we recommended they study. GAO maintains that a more comprehensive review is necessary and suggests that Congress consider directing Interior to (1) convene an independent panel to conduct such a review of the federal

* This is an edited, reformatted and augmented version of U. S. Government Accountability Office publication , GAO-08-691, dated September 2008.

oil and gas fiscal system, and (2) establish procedures to periodically evaluate the state of the fiscal system.

To view the full product, including the scope and methodology, click on GAO-08-691. For more information, contact Frank Rusco at (202) 512-3841 or ruscof@gao.gov.

What GAO Found

In addition to having a low government take, the deep water Gulf of Mexico and other U.S. regions are attractive targets for investment because they have large remaining oil and gas reserves and the U.S. is generally a good place to do business compared to many other countries with comparable oil and gas resources. Multiple studies completed as early as 1994 and as recently as June 2007 indicate that the U.S. government take in the Gulf of Mexico is lower than that of most other fiscal systems. For example, data GAO evaluated from a June 2007 industry consulting firm report indicated that the government take in the deep water U.S. Gulf of Mexico ranked 93rd lowest of 104 oil and gas fiscal systems evaluated. Generally, other measures indicate that the United States is an attractive target for oil and gas investment.

The lack of price flexibility in royalty rates—automatic adjustment of these rates to changes in oil and gas prices or other market conditions—and the inability to change fiscal terms on existing leases have put pressure on Interior and the Congress to change royalty rates in the past on an ad hoc basis with consequences that could amount to billions of dollars of foregone revenue. For example, royalty relief granted on leases issued in the deep water areas of the Gulf of Mexico between 1996 and 2000—a period when oil and gas prices and industry profits were much lower than they are today—could cost the federal government between $21 billion and $53 billion, depending on the outcome of ongoing litigation challenging the authority of Interior to place price thresholds that would remove the royalty relief offered on certain leases. Further, royalty rate increases in 2007 are expected to generate modest increases in federal revenues from future leases offered in the Gulf of Mexico. However, in choosing to increase royalty rates, Interior did not evaluate the entire oil and gas fiscal system to determine whether or not these increases strike the proper balance between the attractiveness of federal leases for investment and appropriate returns to the federal government for oil and gas resources.

Interior does not routinely evaluate the federal oil and gas fiscal system, monitor what other governments or resource owners are receiving for their energy resources, or evaluate and compare the attractiveness of federal lands and waters for oil and gas investment with that of other oil and gas regions. As a result, Interior cannot assess whether or not there is a proper balance between the attractiveness of federal leases for investment and appropriate returns to the federal government for oil and gas resources. Specifically, Interior does not have procedures in place for evaluating the ranking of (1) the federal oil and gas fiscal system or (2) industry rates of return on federal leases against other resource owners. Interior also does not have the authority to alter tax components of the oil and gas fiscal system. All these factors are essential to inform decisions about whether or how to alter the federal oil and gas fiscal system in response to changing market conditions.

September 3, 2008

The Honorable Jeff Bingaman
Chairman, Committee on Energy and Natural Resources
United States Senate

The Honorable Nick J. Rahall II
Chairman, Committee on Natural Resources
House of Representatives

The Honorable Steve Pearce
Ranking Member, Subcommittee on Energy and Mineral Resources
Committee on Natural Resources
House of Representatives

The Honorable Mary L. Landrieu
United States Senate

In fiscal year 2007, domestic and foreign companies received over $75 billion from the sale of oil and gas produced from federal lands and waters, according to the Department of the Interior's (Interior) Minerals Management Service (MMS). The agency further reported that these companies paid the federal government about $9 billion in royalties for such oil and gas development. Clearly, such large and financially significant resources must be carefully developed and managed so that the nation's rising energy needs are met while at the same time the American people are assured of receiving a fair return on publicly owned resources. In May 2007, we reported that, based on studies by industry experts, the amount of money that the U.S. government receives from production of oil and gas on federal lands and waters—the so-called "government take"—was among the lowest in the world. The government take that accrues to any government resource owner is largely determined by the government's oil and gas fiscal system—the precise mix and total amount of payments made to the government for the rights to explore, develop, and sell oil and gas resources. We also noted that several factors needed to be considered to determine whether adjustments to an oil and gas fiscal system are warranted, including the size and availability of the oil and gas resources in place; the costs of finding and developing these resources, including labor costs and the costs of complying with environmental regulations; and the stability of both the oil and gas fiscal system and the country in general.[1] Conceptually, a fair government take would strike a balance between encouraging private companies to invest in the development of oil and gas resources on federal lands and waters while maintaining the public's interest in collecting the appropriate level of revenues from the sale of the public's resources.

Governments and companies negotiate the exploration and development of oil and gas resources under terms of leases or contracts granted by governments. The terms and conditions of such arrangements are established by law and policy, or are negotiated on a case-by-case basis. An important aspect of these arrangements is the oil and gas fiscal system, which defines all applicable payments from the companies to the government resource owners.[2] Oil and gas fiscal systems vary widely across different resource owners, reflecting differences in the mix and weight of the various payments. U.S. federal oil and gas leases operate under a system in which the government transfers title to the oil and gas produced to a company for a period of time, generally in exchange for a lump-sum payment called a bonus bid. The company is then typically subject to the payment of rental rates, royalties, and taxes for any oil and gas that is eventually produced on the lease. In other oil and gas fiscal systems, such as "production sharing" or "profit sharing" systems, the host country and production company enter into a contract to apportion the production or profits between them rather than or in addition to royalty payments.

In recent years, and in response to increasing industry profits and other changing market conditions, many countries have re-evaluated or are re-evaluating their oil and gas fiscal systems. A number of countries have significantly increased their overall government take while others—typically those with marginal or less certain levels of oil and gas resources—have reduced their government take. According to Wood Mackenzie, an energy consulting firm that recently performed a comprehensive study of government take and other measures that determine the attractiveness of different countries to oil and gas investors, the most prominent trend in changing oil and gas fiscal systems has been the imposition of windfall profits taxes or other mechanisms to increase the resource owners' shares of oil and gas revenues from existing projects. Wood Mackenzie estimates that these changes will ultimately result in these countries' collecting additional oil and gas revenues of between $118 billion and $400 billion, depending on future oil and gas prices.[3] For example, the state of Alaska recently increased its government take and changed the terms of contracts to give Alaska larger shares of revenues as oil and gas prices increase. A second trend in changing oil and gas fiscal systems has been an increase in governmental control of resources. For example, Algeria, Russia, and Venezuela have rich resource reserves and have increased the state control over these resources, while increasing their government takes. Other trends include increasing variation of fiscal terms across and within countries to reflect differences in the value of the resources and other factors that affect the attractiveness of these resources to investors. For example, Papua New Guinea and Vietnam are offering terms to encourage production that reflect these countries' status as frontier areas for exploration, while Norway has provided incentives for exploration and continued production on fields with declining production.

A considerable body of legislation governs Interior's authority and obligations to manage resources on federal lands and within federal waters. For example, the Outer Continental Shelf Lands Act (OCSLA) and the Federal Land Policy and Management Act (FLPMA) direct Interior to ensure the United States receives fair market value on the development of its oil and gas resources. In 1976, an Interior report concluded that the government receives "fair market value" when lessees receive no more than a "normal" rate of return. In 1982—the last time Interior convened a task force to comprehensively review its "fair market value" procedures, the task force indicated that fair market value was not the value of the oil and gas eventually discovered or produced; instead it is the value of "the right" to explore and, if there is a discovery, to develop and produce the energy resource. In general, for offshore, Interior has the authority to change most components of the federal oil and gas fiscal system so long as no more than one component is set on automatically adjusting or "flexible" terms, and so long as the Congress does not disapprove the change—pass a resolution of disapproval—within 30 days of receiving notice of Interior's bidding system. However, only the Congress may change the tax components of the oil and gas fiscal system.

To provide more information to the Congress about the nature of the federal oil and gas fiscal system and the attractiveness of the United States as a place in which to invest in oil and gas development, we agreed to build on the information in our May 1, 2007, report, which compared the U.S. government's take with the government takes of other resource owners throughout the world, by reviewing new studies on the subject and adding and updating other information. Specifically, this report (1) evaluates the government take and the attractiveness for investors in the federal oil and gas fiscal system for the Gulf of Mexico and the United States in general, (2) evaluates how the absence of flexibility in this system has led to large foregone revenues from oil and gas production on federal lands and waters as oil and gas prices have risen, and (3) assesses what Interior has done to monitor the performance and appropriateness of the federal oil and gas fiscal system in light of changing market conditions.

To evaluate the attractiveness of the United States for oil and gas investment, we reviewed the results of a study we procured from Wood Mackenzie, a leading industry consultant. In using this study, we reviewed the methodology and controls used by Wood Mackenzie to ensure the accuracy of the data used and the study results. We found the study results and the data that accompanied the study to be sufficiently reliable to meet the objectives of this report.[4] We also evaluated the results of various studies conducted by other industry experts and by MMS, the agency responsible for collecting oil and gas royalties from federal lands and waters. To evaluate the study results, we interviewed study authors and other industry experts to determine the studies' methodologies and the appropriate interpretation of the results. Based on these interviews and our review of the results of the studies, we believe the general approaches taken by the authors of the studies was reasonable and that results of the studies are credible. However, with the exception of the Wood Mackenzie study, we did not fully evaluate each study's methodology or the underlying data used to make the government take estimates. We also purchased and evaluated data from a leading financial firm and evaluated data and information published by the Department of Energy's Energy Information Administration (EIA), the American Petroleum Institute, and other sources. We assessed these data for reliability and deemed them reliable for the purposes of this report. We reviewed academic and government studies that investigated the costs and benefits of various oil and gas fiscal systems, and we interviewed and gathered information from officials from MMS, other governments, and the oil and gas industry. To evaluate how the absence of flexibility in the federal oil and gas fiscal system has led to large foregone revenues from oil and gas production on federal lands and waters as oil and gas prices have risen, we relied on past work evaluating changes in federal oil and gas fiscal terms in the deep water regions of the U.S. Gulf of Mexico. We also evaluated Interior analyses that accompanied recent increases in royalty rates in the U.S. Gulf of Mexico. In addition, we reviewed the Wood Mackenzie study and accompanying data. We also interviewed company officials and industry experts to obtain information on their preferences regarding oil and gas fiscal system characteristics. To assess what Interior has done to monitor the performance and appropriateness of the federal oil and gas fiscal system in light of changing market conditions, we evaluated the extent to which Interior had collected the types of information and done the analysis needed to determine whether or not the oil and gas fiscal system should be changed in light of the recent changes in oil and gas market conditions. To do this, we reviewed Interior studies, policies, and guidance and interviewed officials from MMS; interviewed and collected views from oil and gas companies and industry groups; and evaluated analyses of oil and gas fiscal systems. We neither assessed

Interior's overall management of the federal system, both on and offshore, nor did we attempt to evaluate the costs and benefits of any of Interior's specific changes to the system over time. We conducted this performance audit from May 2007 to September 2008 in accordance with generally accepted government auditing standards. Those standards require that we plan and perform the audit to obtain sufficient, appropriate evidence to provide a reasonable basis for our findings and conclusions based on our audit objectives. We believe that the evidence obtained provides a reasonable basis for our findings and conclusions based on our audit objectives.

RESULTS IN BRIEF

In addition to having a low government take, the U.S. Gulf of Mexico and other U.S. regions are attractive targets for investment because they have large remaining oil and gas reserves and the United States is generally a good place to do business compared to many other countries with comparable oil and gas resources. Multiple studies completed as early as 1994 and as recently as June 2007 indicate that the U.S. government take in the Gulf of Mexico is lower than that of most other oil and gas fiscal systems. For example, data we evaluated from a June 2007 Wood Mackenzie report indicate that the government take in the deep water U.S. Gulf of Mexico ranked as 93rd lowest of 104 oil and gas fiscal systems evaluated. More broadly, other measures indicate that the United States as a whole is an attractive target for oil and gas investment. First, the deep water U.S. Gulf of Mexico and other U.S. oil and gas regions rank high in terms of remaining oil and gas reserves among countries that allow private oil and gas companies to operate on their lands and waters. Second, since 2002 as oil prices have risen and gas prices have remained high by historical standards, the number of oil and gas drilling rigs operating in U.S. lands and waters has increased much faster than in the rest of the world. Specifically, the number of rigs in use globally outside the United States increased by about 18 percent from an annual average of 998 in 2002 to 1,180 through the first 4 months of 2008, while the number of rigs operating in the United States increased by about 113 percent, from an annual average of 831 in 2002 to 1,829 rigs in April 2008. Finally, the United States ranks high among almost all other governments in terms of its general attractiveness for doing business. For example, the World Bank ranked the United States as the third most favorable place to conduct business of 178 countries analyzed in a 2007 study.

The lack of price flexibility in royalty rates and the inability to change fiscal terms for existing leases have put pressure on Interior and the Congress to change royalty rates on future leases in an ad hoc basis with consequences that could amount to billions of dollars of foregone revenue. For example, 1995 legislation granted royalty relief on certain leases issued in the deep water areas of the U.S. Gulf of Mexico between 1996 and 2000—a period when oil and gas prices and industry profits were much lower than they are today—could cost the federal government between $21 billion and $53 billion, depending on the outcome of ongoing litigation concerning the authority of Interior to place price thresholds that would remove the royalty relief offered on certain leases. A royalty relief provision also was included in the Energy Policy Act of 2005 on leases issued during the 5-year period beginning on the date of enactment of this act. Further, two royalty rate increases affecting future U.S. Gulf of Mexico leases were announced by Interior in 2007. These royalty rate increases are expected to generate modest increases in federal revenues from future leases offered in the U.S. Gulf of Mexico. However, in choosing to increase royalty rates Interior did not evaluate the entire oil and gas fiscal system to determine whether or not these increases strike the proper balance between the attractiveness of federal leases for investment and appropriate returns to the federal government for oil and gas resources. As a result, and because the new royalty rates are not flexible with respect to oil and gas prices, Interior and the Congress may again be under pressure from industry or the public to further change royalty rates if and when oil and gas prices either fall or continue rising. Finally, these royalty changes only affect U.S. Gulf of Mexico leases and do not address onshore leases at all, which should also be considered in light of the increases in oil and gas prices. Wood Mackenzie reports that the deep water U.S. Gulf of Mexico ranked in the bottom half of oil and gas fiscal systems in terms of stability based on recent changes to fiscal terms and on the relative lack of built-in flexibility that would allow the fiscal terms to adjust to market conditions. Oil and gas companies we communicated with stated a clear preference for stable fiscal terms, other things being equal. In general, while companies prefer lower government take, it is reasonable to expect that these companies would be willing to pay a higher share of revenues in return for greater assurance that the fiscal terms will not induce balancing changes when market conditions change, such as the windfall profits charges that a number of countries have recently imposed.

Interior does not routinely evaluate the federal oil and gas fiscal system as a whole, monitor what other governments or resource owners worldwide are receiving for their energy resources, or evaluate and compare the attractiveness of the United States for oil and gas investment with that of other oil and gas regions. As a result, Interior cannot assess whether or not there is a proper balance between the attractiveness of federal lands and waters for oil and gas investment and a reasonable assurance that the public is getting an appropriate share of revenues from this investment. Specifically, Interior does not have procedures in place for routinely evaluating the ranking of (1) the federal oil and gas fiscal system or (2) industry rates of return on federal leases against other resource owners. Further, Interior does not have the authority to alter the tax components of the oil and gas fiscal system. All these factors should inform any decisions about whether or how to alter the federal oil and gas fiscal system in response to changing market conditions. While Interior has made many specific changes to components of the oil and gas fiscal system over the years to adjust to changing market conditions, these changes were generally not done as part of a comprehensive review of the system that took into account the relative ranking of the U.S. government take or other comparisons with other countries or regions. In fact, the last time Interior conducted a comprehensive evaluation of the federal oil and gas fiscal system was over 25 years ago. Finally, the lack of a comprehensive re-evaluation of the federal oil and gas fiscal system stands in contrast to the actions of many other governments that have recently re-evaluated or are currently re-evaluating their systems in light of rising oil and gas prices and higher industry profits and rates of return.

Recent large increases in oil and gas prices and industry profits raise obvious questions about whether the public share of oil and gas revenues is appropriate. The fact that the recent studies show that the government take in the deep water U.S. Gulf of Mexico is relatively low and U.S. federal oil and gas regions are attractive places to invest also indicates that the federal oil and gas fiscal system may not strike a proper balance between maintaining competitive investment conditions and providing an appropriate share of revenues to the public from oil and gas sold on public lands and waters. Finally, because Interior has not comprehensively re-evaluated the federal oil and gas fiscal systems for over 25 years, such a comprehensive evaluation of the systems, both on- and offshore, is overdue. Comparing oil and gas fiscal systems and attractiveness for investment is inherently complex and Interior has not collected information needed to perform such a comprehensive review. In the draft report we sent to Interior for comment, we made recommendations to address these issues. In its response, Interior stated that it did not fully concur with our recommendations because it had already contracted for a study that will address many of the issues we raise. However, because Interior's ongoing study is limited in scope and is limited to a specific region in the Gulf of Mexico, rather than a review of the entire federal oil and gas fiscal system as we recommended, we do not find the agency's stated rationale for not agreeing fully with our recommendations to be convincing. Therefore, we believe that Congress may wish to consider directing the Secretary of the Interior to to convene an independent panel to perform a comprehensive review of the federal oil and gas fiscal system. Further, in order to keep abreast of potentially changing market conditions going forward, the Congress may wish to consider directing the Secretary of the Interior to direct the Minerals Management Service and other relevant agencies within Interior to establish procedures for periodically collecting data and information and conducting analyses to determine how the federal government take and the attractiveness for oil and gas investors in each federal oil and gas region compare to those of other resource owners and report this information to the Congress.

BACKGROUND

Interior, created by the Congress in 1849, oversees and manages the nation's publicly owned natural resources, including parks, wildlife habitats, and minerals, including crude oil and natural gas resources, on over 260 million surface acres and 700 million subsurface acres onshore and in the waters of the Outer Continental Shelf. In this capacity, Interior is authorized to lease federal oil and gas resources and to collect the royalties associated with their production, Interior's Bureau of Land Management (BLM) is responsible for leasing federal oil and natural gas resources on land, whereas offshore—including the U.S. Gulf of Mexico—Minerals Management Service (MMS) has the leasing authority. To lease U.S. Gulf of Mexico waters for oil and gas exploration, companies generally must first pay the federal government a sum of money that is determined through a competitive auction and evaluated by Interior against departmental economic and geologic models. This money is called a bonus bid. Companies are required to submit one-fifth of any bid for a lease tract up front at time of bid, and pay the remaining four-fifths balance of their bonus payment and their first year rental payment after acquiring a lease. After the lease is awarded and production begins, the companies must also pay royalties to MMS based on a percentage of the cash value of the oil and gas produced and sold or "in kind," as a percentage of the actual oil or gas produced. Royalty rates for onshore leases are generally 12.5 percent. Royalty rates for leases in the U.S. Gulf of Mexico, prior to 2007, ranged from 12.5 percent for water depths of 400 meters or deeper (referred to as deepwater) to 16-2/3 percent for water depths less than 400 meters (referred to as shallow). In 2007, the Secretary of Interior twice increased the royalty rate for future U.S. Gulf of Mexico leases—in January, the rate for deep water leases was raised to 16-2/3 percent and in October, the rate for all future leases, included those issued in 2008, was raised to 18-3/4 percent.

A considerable body of legislation has been enacted pertaining to the management of resources on federal and Indian trust lands and within federal waters. This legislation includes the Mining Law of 1872, Mineral Lands Leasing Act of 1920, 1947 Mineral Leasing Act for Acquired Lands, Outer Continental Shelf Lands Act of 1953, Federal Land Policy and Management Act of 1976, the Outer Continental Shelf Lands Act Amendments of 1978, Federal Oil and Gas Royalty Management Act of 1982, as well as the Federal Onshore Oil and Gas Leasing Reform Act of 1987, the Outer Continental Shelf Deep Water Royalty Relief Act of 1995, and the Energy Policy Act of 2005. The Outer Continental Shelf Lands Act, as amended (OCSLA) is, among other things, intended to ensure the public "a fair and equitable return" on the resources of the shelf. The law directs the Secretary of Interior to conduct leasing activities to assure receipt of fair market value for the lands leased and the rights conveyed by the federal government. In addition, the Federal Land Policy and Management Act indicated that, unless otherwise provided by statute, it is the policy of the United States to receive "fair market value" for the use of the public lands and their resources. In 1982, Interior's MMS convened a task force to review its fair market value procedures. Upon completion of the task force's work, the Secretary of Interior informed the Congress by letter in March 1983 that the Department had completed its analysis and validation of the process by which it will assure a fair return to the American people. The Secretary indicated in that letter that the process in place will assure the American people a full and fair return as it pertains to bonuses, rentals, royalties, and taxes. The 1983 Interior task force report also provided some clarity regarding a fair return, or fair market value. The report indicated that the market value of a lease is not the market value of the oil and gas eventually discovered or produced. Instead, it is the value of the right to explore and, if there is a discovery, develop and produce the energy resource.

Currently Interior has the legal authority to change most aspects of the oil and gas fiscal system. Specifically, Interior is allowed by statute to change bid terms for offshore leases including the royalty rate, the bonus bid structure, rental terms, and even the minimum 12.5 percent royalty rate, so long as there is only one variable or "flexible" term—such as a royalty rate that adjusts upwards or downwards with oil and gas prices—in the resulting system, and so long as Congress does not pass a resolution of disapproval within 30 days of Interior's changes to the system.[5] With regard to onshore leases, Interior is generally allowed by statute to change bid terms including the royalty rate, the bonus bid structure, rental terms, and the minimum royalty rate so long as the bid structure meets certain bid terms,[6] but with certain additional limits on flexibility than the offshore leases. Over the past 25 years, Interior has implemented several programs that adjusted royalty rates or other system components. Such programs included the net profit share leases, which were for offshore leases that based royalties on a percentage of net profits derived from production; sliding scale royalty rates, which was an onshore royalty rate system based on changing production levels; and royalty rate reduction for stripper wells and lower-grade, more viscous crude oil, where onshore oil wells producing less than 15 barrels of oil per day were eligible for royalty rate reductions. Interior officials told us they also "experimented" with a variety of flexible royalty rate and profit sharing systems in the early 1980s, but found them difficult to administer and validate the amount of payments due to MMS, which in Interior's estimation more than offset any enhanced flexibility associated with a variable royalty rate.

THE GULF OF MEXICO HAS A RELATIVELY LOW U.S. GOVERNMENT TAKE AND THE UNITED STATES IS AN ATTRACTIVE PLACE TO INVEST IN OIL AND GAS DEVELOPMENT

Multiple studies completed as early as 1994 and as recently as June 2007 all indicate that the U.S. government take in the Gulf of Mexico is lower than most other oil and gas fiscal systems. Four recent studies by private consultants or resource owners indicate that the U.S. government take in the Gulf of Mexico is relatively low. For example, data we evaluated from a June 2007 report by Wood Mackenzie reported that the government take in the deep water U.S. Gulf of Mexico ranked as the 93rd lowest out of 104 oil and gas fiscal systems evaluated in the study. Other U.S. oil and gas regions are also listed in the Wood Mackenzie study and some but not all other studies. However, these regions are not uniquely under federal jurisdiction, so a direct comparison of the government take in these other regions cannot be used to isolate the federal oil and gas fiscal system. The results of the four studies are summarized below in table 1.

Table 1. Summary of Four 2007 Studies Comparing Government Take Percentages

Study	**Rank (from highest to lowest) of Gulf of Mexico U.S. government take among oil and gas fiscal systems reviewed in each study**	**Government take percentages**		
		Highest	Lowest	Gulf of Mexico
OUR FAIR SHARE: Report of the Alberta Royalty Review Panel, Sept. 18, 2007 (analysis done by the Alberta Department of Energy).	16/19	77.00	39.00	47.50
Cambridge Energy Research Associates: 2002 vs. 2007.	17/17	95.00	49.00	49.00
Van Meurs Corporation: Comparative Analysis of Fiscal Terms for Alberta Oil Sands and International Heavy and Conventional Oils, May 17, 2007.	25/28	92.00	25.00	47.00
Wood Mackenzie: Government Take: Comparing the Attractiveness and Stability of Global Fiscal Systems, Wood Mackenzie, June 2007.	93/104	98.04	18.05	44.09

Source: GAO analysis of four calendar year 2007 government take studies.

As we reported in May 2007, the results of five other studies completed between 1994 and 2006 had similar findings. The information reported by Wood Mackenzie and other such expert studies are used by resource owners and oil and gas companies alike to aid in making investment or policy decisions and these studies represent the best data on government take available. However, we recognize there are limitations with the government take studies and the relative ranking of government take alone is not sufficient to determine whether the federal government is receiving its fair share of oil and gas revenues. A number of other factors that are not part of the government take determine company decisions of where and how much to invest and how much to pay for access to oil and gas resources. These factors include the relative size of oil and gas resource bases in different regions and the relative costs of developing these resources. Thus government take is a major, but not sole factor in determining the attractiveness of a fiscal system for oil and gas development.

When other factors are taken into consideration, the U.S. Gulf of Mexico is an attractive target for investment because it has large remaining oil and gas reserves and the United States is generally a good place to do business compared to many other countries with comparable oil and gas resources. For example, once reserves that are entirely owned by governments are removed from the analysis, of the 104 remaining fiscal regimes ranked by Wood Mackenzie that allow some participation by international oil companies and that have remaining oil and gas reserves, the deep water U.S. Gulf of Mexico ranked 18th highest in terms of remaining oil and gas reserves. Three other U.S. regions were ranked in the top 18 in terms of reserves. These were the U.S. Rocky Mountains (8th), Alaska (14th), and U.S. Gulf Coast (15th), but these regions are not uniquely covered by the federal fiscal regimes, as state and private resource owners may also exist.

Wood Mackenzie also ranked oil and gas fiscal regimes in terms of their attractiveness for investment. Wood Mackenzie's measure of oil and gas fiscal attractiveness took into account both reward associated with factors such as resource size; and risk, including the extent to which government take includes bonuses. With respect to reward, Wood Mackenzie compared the levels of government take with the size of oil and gas fields governed by the various oil and gas fiscal systems. The risk ranking reflected whether or not the system included bonus payments, which increase the risk to investors because they must be paid whether or not economic volumes of oil and gas are eventually found on an oil and gas tract.[7] Risk also included a measure of the extent to which and the way in which the resource owner held an equity share in the resources being developed. The impact of the fiscal terms on the rewards and risks associated with a wide range of hypothetical new investments were assessed under the terms of each of the 103 oil and gas fiscal systems included in this section of the study. Based on these assessments, Wood Mackenzie ranked the deep water U.S. Gulf of Mexico fiscal system as more attractive for investment than 60 (about 58 percent) of the 103 fiscal systems ranked.

More broadly, other measures indicate that the United States is an attractive place to invest in oil and gas production. For example, since 2002 as oil prices have risen and gas prices have remained high by historical standards, the number of oil and gas drilling rigs operating in the United States has increased much faster than in the rest of the world, which indicates companies in recent years have continued to find the United States a conducive place to invest in oil and gas production. Specifically, according to data on crude oil rig counts from Baker Hughes, the number of rigs in use globally excluding the United States increased by about 18 percent from an annual average in 2002 of 998 to 1,180 through the first 4 months of 2008, while the number of rigs operating in the United States increased about 113 percent, from 831 rigs in 2002 to 1,768 rigs in 2007 and 1,829 rigs in April 2008. These increases coincided with the increase in oil and gas prices over the same period and indicate that the United States has remained an attractive place to invest in oil and gas as prices have risen.

While rig counts can reasonably be associated with the attractiveness of a region for development and production, they do not tell the whole story. For example, according to Baker Hughes, the Gulf of Mexico rig count fluctuated over the longer term and has decreased in recent years. Specifically, from 1973 to 1981, rig counts in the Gulf of Mexico increased, from 80 to 231, before generally decreasing to 45 in 1992. They then generally rose again until 2001. From 2001 to April 2008, the annual rig count in the Gulf of Mexico decreased from 148 to 58. This decline has occurred despite the Gulf of Mexico being generally considered an attractive target for investment, both from the perspective of the government take and because of the potential for significant oil and gas resources.

Other analyses report the oil and gas industry appears to have performed favorably in recent years compared with other industries.

- The Energy Information Administration reported in December 2007 that from 2000 through 2006, the return on equity, which compares a company's profit with the value of the shares held by the company's owners, for the major energy producers, referred to as Financial Reporting System (FRS) companies, averaged 7 percentage points higher than that of the U.S. Census Bureau's "All Manufacturing Companies."[8] According to the report, this reversed a trend where the return on equity for the major energy producers averaged 2 percentage points lower than All Manufacturing Companies from 1985 to 1999.
- The American Petroleum Institute, in a 2007 study, showed that from 2000 to 2005, the average return on investment for oil and gas production was about 61 percent higher than for the Standard & Poor's (S&P) industries.[9] However, the average return on investment for the industry has matched or exceeded the returns for the S&P industrials only in recent years; over the 25-year period from 1980 to 2005, the average return on investment for oil and gas production was about 18 percent lower than for the S&P industries.
- A GAO analysis found that the "upstream," or exploration and production segments, of the domestic oil and gas production companies also received higher rates of return than companies operating in other U.S. manufacturing industries from 2002 through 2006. We analyzed financial data from S&P's Compustat and EIA's FRS. From 2002 through 2006, the upstream segments of the domestic oil and gas production companies have averaged a 17.4 percent return on investment,[10] compared with 15.2 percent for all other manufacturing companies.[11] When both upstream and "downstream" the refining and marketing segments are included in the analysis, the oil and gas industry return on investment averaged over 20 percent during this period. This short term picture, however, contrasts with a longer-term analysis, which shows the oil and gas industry receiving a return on investment that is comparable, or slightly lower, than that received by other manufacturing industries over the past 30 years. Our analysis found that during this period, upstream oil and gas production has averaged 11.2 percent return on investment with the entire oil and

gas industry receiving an average 13.7 percent return on investment. All other manufacturing companies have averaged 12.3 percent return on investment during this period. This recent improvement in financial performance from 2000 through 2006 coincided with rising oil and gas prices. Further, since 2006, oil and gas prices have risen even higher, while EIA's most recent projections to 2030 are for oil and gas prices to remain much higher than they were for most of the period 1985 through 1999. In addition, the United States is also generally ranked favorably as a place to conduct business by the World Bank and by business media sources, including *The Economist, and AM Best*. For example, the World Bank ranked the United States as the third most favorable place to conduct business of 178 countries analyzed in a 2007 study.[12] *The Economist* in October 2007 ranked the United States as the ninth highest of 82 countries analyzed for projected favorability of business environment from 2008 to 2012. Finally, as of February 2008, the United States remained in the top tier—of five possible tiers—on *AM Best's* countries for business risk index, meaning the United States generally posed the least risk for investors of the five possible levels assigned.

The Inflexibility of Royalty Rates to Changing Oil and Gas Prices Has Cost the Federal Government Billions of Dollars in Foregone Revenues

The lack of price flexibility in royalty rates and the inability to change fiscal terms for existing leases have put pressure on Interior and the Congress to change royalty rates in the past on future leases on an ad hoc basis. For example, in 1980, a time when oil prices were comparable in inflation-adjusted terms to today's prices, Congress passed a windfall profit tax, which amounted to an excise tax per barrel of oil produced in the United States. Congress repealed that tax in 1988 at time when oil prices had fallen significantly from their 1980 level. The tax attempted to recoup for the federal government much of the revenue that would have otherwise gone to the oil industry as a result of the decontrol of oil prices.

Further, in 1995—a period with relatively low oil and gas prices—the federal government enacted the Outer Continental Shelf Deep Water Royalty Relief Act (DWRRA). In implementing the DWRRA for leases sold in 1996, 1997, and 2000, MMS specified that royalty relief would be applicable only if oil and gas prices were below certain levels, known as "price thresholds," with the intention of protecting the government's royalty interests if oil and gas prices increased significantly. MMS did not include these same price thresholds for leases it issued in 1998 and 1999. In addition, the Kerr-McGee Corporation—which was active in the Gulf of Mexico and is now owned by Anadarko Petroleum Corporation—filed suit challenging Interior's authority to include price thresholds in DWRRA leases issued from 1996 through 2000. Recently, the U.S. District Court for the Western District of Louisiana granted summary judgment in favor of Kerr-McGee concerning the application of price thresholds to those leases and this ruling is currently under appeal.[13] In our June 2008 report on the potential foregone revenues at stake in the Kerr-McGee litigation, we found that the value of future forgone royalties is highly dependent upon oil and gas prices, and on production levels.[14] Assuming that the District Court's ruling is upheld, future foregone royalties from all the DWRRA leases issued from 1996 through 2000 could range widely—from a low of about $21 billion to a high of $53 billion,[15] depending on the outcome of ongoing litigation concerning the authority of Interior to place price thresholds that would remove the royalty relief offered on certain leases. The $21 billion figure assumes relatively low production levels and oil and gas prices that average $70 per barrel and $6.50 per thousand cubic feet over the lives of the leases. The $53 billion figure assumes relatively high production levels and oil and gas prices that average $100 per barrel and $8 per thousand cubic feet over the lives of the leases. A royalty relief provision was also included in the Energy Policy Act of 2005 on leases issued during the 5-year period beginning on August 8, 2005.

In 2007, the Secretary of the Interior twice increased the royalty rate for future Gulf of Mexico leases—in January, the rate for deep water leases was raised to 16-2/3 percent and in October, the rate for all future leases in the Gulf, including those issued in 2008, was raised to 18-3/4 percent. Interior estimated these actions will increase federal oil and gas revenues by $8.8 billion over the next 30 years. The January 2007 increase applied only to deep water Gulf of Mexico leases; the October 2007 increase applied to all water depths in the Gulf of Mexico. These royalty rate increases appear to be a response by Interior to the high prices of oil and gas that have led to record industry profits and raised questions about whether the existing federal oil and gas fiscal system gives the public an appropriate share of revenues from oil and gas produced on federal lands and waters. However, the royalty rate increases do not address these record industry profits from existing leases at all and high profits will likely remain as long as the existing leases produce oil and gas or until oil and gas prices fall. In addition, in choosing to increase royalty rates, Interior did not evaluate the entire oil and gas fiscal system to determine whether or not these increases were sufficient to balance investment attractiveness and appropriate returns to the federal government for oil and gas resources. On the other hand, according to Interior, it did consider factors such as industry costs for outer continental shelf exploration and development, tax rates, rental rates, and expected bonus bids. Further, because the new royalty rates are not flexible with respect to oil and gas prices, Interior and the Congress may again be under pressure from industry or the public to further change the royalty rates if and when oil and gas prices either fall or continue rising. Finally, these royalty changes only affect Gulf of Mexico leases and do not address onshore leases at all, which should also be considered in light of the increases in oil and gas prices.

In addition, Wood Mackenzie reports that the deep water U.S. Gulf of Mexico ranked in the bottom half of countries in terms of oil and gas fiscal system stability based on repeated changes to fiscal terms for future leases and on the relative lack of built-in flexibility that would allow the fiscal terms to adjust to market conditions. Specifically, the Wood Mackenzie study ranked the deep water U.S. Gulf of Mexico fiscal terms as lower than 71 (about 72 percent) of the 103 oil and gas fiscal systems. In contrast, among the key trends among governments in recent years has been to make fiscal terms more responsive to market conditions. By adding such progressive features to oil and gas fiscal systems including royalty rates that increase with oil and gas prices, these other entities are making their systems more stable over time by reducing incentives for industry or the public to push for ad hoc changes in fiscal terms as future prices change. Wood Mackenzie's measure of fiscal stability combines two criteria: recent history of changes to fiscal terms and built-in flexibility. As discussed previously in this report, changes to royalty rates occurred in the Gulf of Mexico three times since 1995, with the royalty relief in the mid 1990s and the two increases in royalty rates in 2007. However, as noted above, the study was conducted before the second 2007 increase in royalty rates, so the government take would likely have increased but the U.S. stability rating could have fallen in the intervening period. Built-in flexibility reflects the relative degree to which a fiscal system is regressive or progressive, with more progressive systems being more flexible. A flexible system does not mean changing the fiscal terms of existing contracts but having a system in place that automatically adjusts to changing economic and market conditions.

Oil and gas companies we communicated with stated a clear preference for stable fiscal terms, other things being equal. Overall, oil and gas companies may be more willing to invest in flexible systems, given that they tend to be inherently more stable and therefore are less likely to be arbitrarily changed on a recurring basis. Oil and gas companies and industry trade associations we contacted provided us a range of views on the advantages and disadvantages of various oil and gas fiscal systems, and generally indicated that one of the most important features of any system is its stability and predictability. Stability of fiscal terms is important because oil and gas companies are making very long-term investments and uncertainty about whether or not the resource owner will change the fiscal terms during the lifetime of the investment adds to the investment risk. The respondents also said that the terms of the oil and gas fiscal system should consider industry exploration and development costs, the likelihood of discovery, and political and economic risks. While companies surely prefer lower government take, all else constant, to the extent that stability is also preferred, a more stable system may be able to remain competitive for investment while resulting in a higher government take than a less stable system. In particular, companies may be willing to pay a larger average share of oil and gas revenues if they believe that oil and gas fiscal systems will not change when market conditions change, such as the windfall profits charges that a number of countries have recently imposed. Such willingness to accept lower expected profits in exchange for lower risk is a common feature of investment markets.

In addition to the potential trade-off between oil and gas fiscal system stability and government take, companies may be willing to pay higher average shares of revenues if governments bear some of the risk that companies take on when they purchase the rights to explore for oil and gas. For example, in the United States as well as for a number of other governments, leases are awarded through a bidding process that requires companies to pay bonus bids for the rights to explore and develop leases. With regard to bonus bids, there are advantages to requiring such bids. First, when companies have to compete with one another to win a lease, the lease is more likely to be awarded to a company with the expertise and resources to properly explore and develop the resources on the lease than if leases are awarded using some other rationing mechanism that does not take into account how much companies are willing to pay for the lease. In addition, it guarantees the public some revenue early on in the exploration and development process, which can take a number of years to complete. However, the use of bonus bids pushes a great deal of risk onto oil and gas companies and requires them to estimate many uncertain factors, including the amounts of oil and gas that will ultimately be produced on the lease, the costs of that production, and the prices of gas and oil over the entire working life of the lease. In general, by increasing the risk that companies bear, these companies will have to expect to receive a higher rate of return to be willing to take on the project. In fiscal systems requiring bonus bids or other up-front payments, the companies bear the risk that leases will not generate economically significant oil and gas production. In fact, in the United States, a large proportion of leases that companies have paid for do not generate economic levels of production and the companies, after purchasing the lease, and paying rent for the duration of the initial term of the lease and whatever resources they spent on exploring for oil and gas, simply let the lease revert back to the government when the initial term expires.

Some oil and gas fiscal systems mitigate the risk associated with up-front company expenditures by allowing the companies to recover exploration and development costs prior to starting higher royalty payments. For example, Alberta, Canada, has used such fiscal terms. Other fiscal systems share risk with companies by more strongly linking government take to company profits. In such oil and gas fiscal systems, government take is low in early years of a lease, when exploration and early development are being undertaken, but increases if production increases or if oil and gas prices increase once production begins. The state of Alaska has recently changed its fiscal terms to increase its government take and to increase the linkage between government take and company profits. Both Alberta and the state of Alaska have higher government takes than the U.S. Gulf of Mexico according to the Wood Mackenzie study.

Interior Does Not Have a System in Place to Evaluate Whether the Federal Fiscal System Is in Need of Reassessment

Interior does not routinely evaluate the federal oil and gas fiscal system as a whole, monitor what other resource owners worldwide are receiving for their energy resources, or evaluate and compare the attractiveness of the United States for oil and gas investment with that of other oil and gas regions. As a result, Interior cannot assess whether or not there is a proper balance between the attractiveness of federal lands and waters for oil and gas investment and a reasonable assurance that the public is getting an appropriate share of revenues from this investment. This is true of the U.S. Gulf of Mexico as well as other federal oil and gas producing regions. Interior does not have procedures in place for routinely evaluating the ranking of (1) the federal oil and gas fiscal system against other resource owners or (2) industry rates of return on federal leases compared to other U.S. industries which could factor into any decisions about whether or how to alter the fiscal systems in response to changing market conditions. Interior officials told us that they have a "bid adequacy review process" for offshore leases that determines whether the bonus bid meets criteria designed to ensure fair market value of the leased tract but that onshore leases do not have a similar bid adequacy provision. Moreover, Interior maintains it has been responsive to changes in market conditions through revisions to lease terms, including changes in minimum bonus bid levels, fluctuating royalty rates, and price thresholds. However, as we have discussed previously in this report, bonus bids have both positive and negative sides with respect to their likely impact on overall government take. Further, frequent adjustments to fiscal terms are not looked on favorably by industry, especially when they involve increases in royalty rates or other charges. Interior indicated, in commenting on the report draft, that it is in the process of evaluating other fiscal approaches such as sliding scale royalties for some oil and gas leases.

We did not evaluate the effectiveness of the bid adequacy review process in terms of its intended goal of ensuring bonus bids on offshore federal leases are competitive. However, even assuming that these bids are competitive, we do not think that this is sufficient to ensure that the other elements of the system are appropriately balancing the interests of taxpayers and oil and gas companies. In light of the complexity of oil and gas fiscal systems, the great deal of uncertainty surrounding the volumes and future prices of oil and gas, and the costs of producing it, oil and gas companies cannot be expected to accurately forecast all the factors that will ultimately determine the value of a lease at the time that lease is sold. As a result, oil and gas company profits have tended to rise and fall over time with oil and gas prices, putting pressure on Interior to alter fiscal terms in a reactive rather than a strategic way. Further, the fact that Interior does not apply the same or a similar bid adequacy process for onshore leases raises questions about how Interior, overall, is providing reasonable assurance that even the bonus bids it receives are competitively determined in all publicly owned oil and gas producing regions.

While Interior has made many specific changes to components of the federal oil and gas fiscal system over the years to adjust to changing market conditions, these changes were generally not done as part of a comprehensive review of the fiscal system that took into account the relative ranking of the U.S. government take or other comparisons with other countries or regions. The last time Interior conducted a comprehensive evaluation of the oil and gas fiscal system was over 25 years ago. The lack of a recent comprehensive re-evaluation of the U.S. federal fiscal system stands in contrast to the actions of many other governments that have recently reevaluated or are currently re-evaluating their fiscal systems in light of rising oil and gas prices and higher industry profits and rates of return. For example, as previously discussed in this report, a number of countries have recently imposed windfall profits taxes or other mechanisms to increase the resource owners' shares of oil and gas revenues from existing projects. Wood Mackenzie estimates that these changes will ultimately result in these countries' collecting additional oil and gas revenues of between $118 billion and $400 billion, depending on future oil and gas prices.

In evaluating an oil and gas fiscal system, all components of the system, including bonus bids, land rental rates, royalties, and oil and gas company taxes, must be considered. However, while Interior has a great deal of expertise and data from years of administering and collecting revenues from oil and gas leases on federal lands and waters that would be essential for any review of the federal oil and gas fiscal system, they do not have the authority to change taxes and, therefore, cannot fully revise the system without legislative action by the Congress. Further, it is essential to keep federal leases competitive with other potential investments governed by different fiscal systems. Therefore, in addition to input from Interior, oil and gas industry experts must also be consulted in any comprehensive review of the federal oil and gas fiscal system. For example, when Alberta recently reviewed its oil and gas fiscal system, it convened a panel that included experts from academia, energy research and consulting firms, and the energy industry and also hired a consultant to evaluate the system and make specific recommendations. Following this review, Alberta increased some elements of the oil and gas fiscal system. However, prior to this review, Canadian government corporate taxes were reduced, which made Alberta more attractive for investors. In any comprehensive review of the U.S. oil and gas fiscal system, taxes may need to be part of the discussion. Therefore, congressional action may be needed to change the federal oil and gas fiscal system, if changes are ultimately determined to be appropriate.

CONCLUSIONS

Oil prices have increased in recent years to levels not seen since the late 1970s and early 1980s when adjusted for inflation. Natural gas prices have also been high by historical standards in recent years. These high prices have coincided with rising oil company profits. Moreover, the EIA's long-term outlook projects these prices to remain much higher than what they had been for much of the past 25 years. Our work indicates that federal oil and gas leases in the deep water U.S. Gulf of Mexico and other U.S. regions are attractive investments and that the government take in the U.S. Gulf of Mexico ranks among the lowest across a large number of other oil and gas fiscal systems. Our work further indicates that other measures, including fiscal attractiveness and rates of return, indicate the U.S. Gulf of Mexico and other U.S. oil and gas producing regions are attractive places to invest. However, the regressive nature of the U.S. federal fiscal systems and other factors have caused these fiscal systems to be unstable over time and this adds risk to oil and gas investments and may reduce the amount oil and gas companies are willing to pay in total for the rights to explore and develop federal leases. Because of these facts and because Interior has not re-evaluated its oil and gas fiscal system in over 25 years, a comprehensive re-evaluation is called for. While Interior could collect data and commission studies to re-evaluate the federal fiscal system, the agency does not currently have the information to fully compare the federal fiscal system with those of other governments, including states or foreign countries. In addition, Interior does not have the authority to make changes to all elements of federal fiscal system if such changes were found to be desirable. Finally, because of the complexity of evaluating oil and gas fiscal systems and the importance of striking a balance between remaining an attractive place for investment and providing revenue to the federal government, it is important that independent experts also be consulted as well as representatives from the oil and gas industry.

MATTER FOR CONGRESSIONAL CONSIDERATION

In the draft report we sent to Interior for comment, we made recommendations to address these issues. In its response, Interior stated that it did not fully concur with our recommendations because it had already contracted for a study that will address many of the issues we raise. However, because Interior's ongoing study is limited in scope and is limited to a specific region in the Gulf of Mexico, rather than a review of the entire federal oil and gas fiscal system as we recommended, we do not find the agency's stated rationale for not agreeing fully with our recommendations to be convincing. Therefore, we believe that Congress may wish to consider directing the Secretary of the Interior to convene an independent panel to perform a comprehensive review of the federal oil and gas fiscal system.

Further, in order to keep abreast of potentially changing market conditions going forward, the Congress may wish to consider directing the Secretary of the Interior to direct the Minerals Management Service and other relevant agencies within Interior to establish procedures for periodically collecting data and information and conducting analyses to determine how the federal government take and the attractiveness for oil and gas investors in each federal oil and gas region compare to those of other resource owners and report this information to the Congress.

AGENCY COMMENTS AND OUR EVALUATION

The Department of the Interior provided us comments on a draft of the report. Overall, the department agreed that it is important to reassess the federal oil and gas fiscal system but did not fully concur with either of our two recommendations to (1) perform a comprehensive review of the system using an independent panel and (2) adopt policies and procedures to keep abreast of important changes in the oil and gas market and in other countries' efforts to adjust their oil and gas management practices in light of these changes. We disagree with Interior's rationale for its lack of full concurrence with our recommendations and have, therefore, elected to reframe the recommendations into Matters for Congressional Consideration in the final report.

In response to our first recommendation, Interior indicated that it would be premature and duplicative for the department to undertake such a review because it had recently contracted with an outside party to conduct a 2-year study of the policies affecting the pace of area-wide leasing and revenues in the Central and Western Gulf of Mexico. We disagree that our recommended review is either premature or duplicative with this Interior study effort. First, a comprehensive review is overdue, given that Interior has not performed a comprehensive evaluation of the oil and gas fiscal system in over 25 years and in light of the dramatic increases in oil and gas prices and industry profits in recent years. Further, as documented in this report, many other oil and gas owners have been re-evaluating and changing their oil and gas fiscal systems in response to these recent market conditions. The Congress and the public are justifiably concerned about whether the federal government is getting a fair return for its energy resources as oil and gas company profits have reached record levels. In addition, our recommended review would not be duplicative with Interior's ongoing study, which is geographically limited to only two sections of the Gulf of Mexico. In contrast, we recommended that Interior review all its oil and gas fiscal systems, both onshore and offshore. Nor does Interior's ongoing study cover the full scope of review that we recommended, including looking at how other resource owners are managing their oil and gas fiscal systems. Further, Interior's ongoing study does not explicitly look at the stability of the system as we recommended and this appears to be a critical factor influencing changes to oil and gas fiscal systems globally. Finally, Interior's ongoing effort does not utilize an independent panel. We believe it is essential to empanel an independent body, representative of major stakeholders, including those representing the interests of industry and the public, in order to develop recommendations that strike an appropriate balance between remaining and attractive place for investment and providing revenue to the federal government.

In response to our second recommendation, Interior implied that such an effort was unnecessary because Interior agencies that lease federal minerals already keep abreast of current literature on fiscal systems of other resource owners. During our work, we identified only one Interior study done over the past 25 years that provided information on the U.S. government take compared to other fiscal systems. While that one Interior study issued in 2006 showed, similar to our work, that the U.S. government take was low compared to other fiscal systems, it is also worth noting that the study itself relied on dated 1994 government-take information. Therefore, we do not believe that Interior has adequately kept abreast of important trends in oil and gas management, especially as it relates to how other resource owners are managing these resources. In addition, our recommendation went further than simply keeping abreast of current literature. In particular, our recommendation sought to have Interior monitor and report on how the federal government's fiscal terms for oil and gas development compare with the terms of other resource owners worldwide.

Interior's full letter commenting on the draft report is printed as appendix III, and our detailed response follows. In addition, Interior made technical comments that we have addressed as appropriate.

Frank W. Rusco
Acting Director,
Natural Resources and Environment

APPENDIX I: SCOPE AND METHODOLOGY

We performed our work at the Department of Interior's (Interior), Bureau of Land Management's (BLM), and Minerals Management Service's (MMS) offices and in Washington, D.C. from May 2007 to September 2008 in accordance with generally accepted government auditing standards. We focused our analysis of government take and industry rates of return on the U.S. Gulf of Mexico because it represents approximately 79 percent of oil and 50 percent of gas production on federal leases, and because there are complicating factors for onshore oil and gas leases, such as state and local taxes or fees that may differ by locality, which the available studies do not fully address. We did evaluate information that applied more broadly to the United States, specifically with respect to overall measures of the attractiveness of the United States for oil and gas investment. However, we cannot infer from our review of the Gulf of Mexico federal oil and gas leases how the data on federal government take or industry returns to investment are applicable to federal onshore leases. In general, the results of this review can compare the federal system associated with the U.S. Gulf of Mexico to that of other oil and gas fiscal systems but cannot provide specific prescriptive recommendations for how to change the federal fiscal system to achieve a fair return for the public from sale of oil and gas on public lands and waters. We also compared the federal oil and gas fiscal system to all types of fiscal systems around the world to encompass the range of choices that oil and gas companies are faced with when deciding where to invest.

To determine the degree to which the federal government is receiving a fair return, our work included reviewing various pieces of energy resource management legislation enacted over the last several decades. This included, among others, the Outer Continental Shelf Lands Act of 1953 (OCSLA) and its amendments and the Federal Land Policy and Management Act of 1976 (FLPMA) and its amendments. We also collected and analyzed various pieces of Interior energy resource policy and management information. To evaluate how the U.S. government take compares to those in other countries, we reviewed the results of a study procured from Wood Mackenzie, a leading industry consultant, and recent studies conducted by other private consultants or resource owners. We also collected and analyzed various studies generated by MMS, the agency responsible for collecting oil and gas royalties from federal lands and waters and interviewed private consulting firm officials. In evaluating the study results, we conducted interviews with study authors and an industry expert to discuss the study methodologies and the appropriate interpretation of the results. Based on these interviews and our review of study results, we believe the general approach that these study authors took was reasonable and that the results of the studies are credible. However, we did not fully evaluate each study's methodology or the underlying data used to make the government take estimates. Overall, because all the studies came to similar conclusions with regard to the relative ranking of the U.S. federal government, and because such studies are used by oil and gas industry companies and governments alike for the purposes of evaluating the relative competitiveness of specific oil and gas fiscal systems, we are confident that the broad conclusions of the studies are valid. To assess the extent to which the United States' oil and gas fiscal system is able to remain stable as market conditions change, we relied heavily on the study and data we obtained from Wood Mackenzie. We interviewed industry experts and gathered information regarding the types of fiscal systems and the relative stability offered with each. We interviewed company officials and industry experts to obtain information on their preferences regarding fiscal system characteristics.

We also purchased data from Compustat and analyzed that data and data published by the Energy Information Administration. The financial data we procured are widely used by private companies and governments for purposes of comparing company and industry rate of return over time, because Interior in the past used rate of return as a credible measure to evaluate the profitability of the Gulf of Mexico for firms conducting oil and gas exploration there versus the relative profitability of other manufacturing firms operating in the United States. We also evaluated data reported by the American Petroleum Institute and other sources. Further, we reviewed various reports prepared over the last 2 years by private sources on the profitability of oil and gas companies operating in the U.S. versus operating elsewhere in the world. We also spoke to industry officials regarding aspects of the various fiscal systems in which they operate. Finally, we discussed the issue of a "fair return" with various Interior, BLM, and MMS officials, as well as members of the oil and gas industry. To determine what steps Interior takes to get reasonable assurance that the federal government take provides a fair return to the public, we reviewed Interior studies and procedures, and interviewed officials from MMS.

We conducted this performance audit from May 2007 to September 2008, in accordance with generally accepted government auditing standards. Those standards require that we plan and perform the audit to obtain sufficient, appropriate evidence to provide a reasonable basis for our findings and conclusions based on our audit objectives. We believe that the evidence obtained provides a reasonable basis for our findings and conclusions based on our audit objectives.

APPENDIX II: COMPANIES RECEIVING ROYALTY RELIEF IN THE U.S. GULF OF MEXICO

According to Interior, companies operating in the U.S. Gulf of Mexico had received more than $1.3 billion in royalty relief through September 30, 2007. Table 2 lists the companies that have received royalty relief under DWWRA and the amounts of that relief. Six companies had signed agreements with Interior, allowing thresholds to be placed for royalties to be paid in the future. Those companies are BP Exploration and Production; ConocoPhillips & Burlington Resources Offshore, Inc.; Marathon; Shell; Walter Hydrocarbons; and Walter Oil and Gas. According to Interior information dated February 4, 2008, ConocoPhillips & Burlington Resources Offshore, Inc., had not received royalty relief.

Table 2. Amounts of Royalty Relief Received by Companies

Company	Ownership	Royalty Relief Received to Date	Signed agreement with Interior to include price thresholds
ATP Oil & Gas Corporation	Public – United States	7,080,958	No
BHPBilliton	Public – Australia	6,480,679	No
BP Exploration & Production	Public – United Kingdom	172,508,633	Yes
Chevron USA/Texaco/Union Oil	Public – United States	4,003,495	No
Devon Energy Corporation/ Ocean/Santa Fe	Public – United States	143,808,801	No
Dominion Exploration	Public – Italy	126,504,055	No
EnCana Gulf of Mexico	Public – Canada	43,908	No
ENI Deepwater	Public – Italy	27,176,887	No
Howell Group	Public – United States	46,867	No
Anadarko Petroleum Corporation/Kerr McGee/Offshore Shelf/Westport	Public – United States	142,406,788	No
Marathon Oil Corporation	Public – United States	1,393,586	Yes
Mariner Energy, Inc.	Public – United States	44,050,427	No
Marubeni	Public – Japan	26,477,247	No
Newfield Exploration Corp.	Public – United States	10,338,890	No
Nexen Inc.	Public – Canada	129,518,866	No
NI Energy Venture	Public – Japan	406,747	No
Nippon Oil Exploration	Public – Japan	22,897,836	No
Noble Corp.	Public – Cayman Islands	1,137,105	No
Occidental Petroleum Corp.	Public – United States	109,653,662	No
Petrobras America	Semipublic – Brazil	13,354,061	No
Pioneer Natural Resources Co.	Public – United States	128,068,000	No
Pogo Producing Co.	Public – United States	7,414,106	No
Royal Dutch Shell	Public – Netherlands	27,399,688	Yes
Total E&P	Public – France	171,648,800	No
Walter Oil & Gas Corp./ Walter Hydrocarbons	Private – United States	1,286,768	Yes
Total		**1,325,106,861**	

Source: GAO analysis of Interior data dated February 4, 2008.

Note: Numbers do not add exactly due to rounding.

APPENDIX III: COMMENTS FROM DEPARTMENT OF THE INTERIOR

Note: GAO comments supplementing those in the report text appear at the end of this appendix.

United States Department of the Interior

OFFICE OF THE SECRETARY
Washington, D.C. 20240

AUG - 8 2008

The Honorable Frank Rusco
Director, Government Accountability Office
441 G Street, NW
Washington, D.C. 20548

Dear Mr. Rusco:

Thank you for the opportunity to review and comment on the Government Accountability Office draft report "Oil and Gas Royalties: The Federal System for Collecting Oil and Gas Revenues Needs Comprehensive Reassessment" (GAO-08-691). We appreciate the efforts of the GAO and have consistently worked closely with the GAO on this and previous reports over the years. We agree that it is important to review the Federal oil and gas fiscal system, but we do not fully concur with the recommendations in your draft report. The GAO's recommendations to conduct more reviews attest to the complexity of evaluating the Federal oil and gas fiscal system.

Background

In June 2006, Congress asked the GAO to review royalties on Federal oil and gas production and provide advice on what changes if any should be made to ensure that the public is receiving fair value. The GAO lists three objectives in its draft report: 1) evaluate the attractiveness for oil and gas investors of the Federal oil and gas fiscal system; 2) evaluate how the absence of flexibility in this system has led to large forgone revenues from oil and gas production on Federal lands and waters; and 3) assess what the Department of the Interior (Interior) has done to monitor the performance and appropriateness of the Federal oil and gas fiscal system. As noted in the draft report, comparing fiscal systems for oil and gas is inherently complex. The Minerals Management Service agrees with the implications of the GAO report that we should evaluate fiscal system designs that are responsive to market conditions and keep informed of what other countries are doing in their fiscal systems; however, we have concerns with other findings and statements in the GAO draft report, which are discussed below.

Areas of Concern

See comment 1.

In addressing the first objective noted above, the GAO draft report relies heavily on measures of Government take, but does not clarify the link between Government take and investment attractiveness to investors, nor does the draft report relate the significance of Government take to the purposes of the Outer Continental Shelf Lands Act and the Federal Land Policy and Management Act.

See comment 2.

The second objective of the GAO study appears to lead to a predetermined conclusion. The report's conclusion that inflexibility in the system is responsible for significant reductions in Federal receipts is not supported by the GAO's analysis presented in the draft report. The example cited by the GAO concerns the deep water leases issued in 1998 and 1999. At the time, the MMS was implementing the special requirements of the Deep Water Royalty Relief Act. The MMS acted in a flexible and creative way to set the elements of the fiscal terms of its leases in a manner consistent with anticipated technical, as well as market, conditions at time of sale. The GAO also states that the MMS did not evaluate the entire oil and gas fiscal system to determine whether or not the two royalty increases provided the proper balance between the attractiveness of Federal leases for investment and appropriate returns to the Federal Government. Interior did evaluate components of the Federal leasing system, including tax and production losses from imposing the resulting royalty increases that, in large part, were a prudent and incremental response to emerging market conditions.

See comment 3.

The GAO does not credit the MMS for efforts related to the third objective, which is monitoring the performance of the Federal fiscal system. Even though Interior may not have conducted a "comprehensive evaluation" of the Federal oil and gas fiscal system as defined by the GAO, the MMS evaluates the expected OCS resources and market conditions in light of our organic statutes and tax laws when setting the fiscal terms for each lease sale and analyzing the results following each sale.

See comment 4.

Onshore Federal royalty issues, while not directly addressed by the GAO in this report, differ from offshore due to the intermingling of Federal, State, Indian, and private lands. The Federal take is severely limited by the revenue split with States. Bonus money paid for onshore leasing is, by law, the result of competitive bidding with a minimum acceptable bid set by Congress of not less than $2.00 per acre. Multiple bids received at onshore lease auctions are an indication of market value, with the high bidder generally being awarded the lease. The price of natural gas is set by the domestic market, and, generally, higher royalty rates will be passed on to consumers as higher gas prices.

See comment 5.

Senior Departmental officials have consistently interpreted the mandates of the OCSLA as receipt of fair market value, expeditious exploration and development, encouragement of competition, protection of human health and the environment, resource conservation, etc.—not maximization of Government receipts as the GAO implies. We believe Interior bureaus have succeeded in achieving the purposes of the OCSLA and FLPMA. In its oversight role, Congress should hold Interior to the standards encompassed in existing statutes. Maximizing Federal mineral revenues or achieving some worldwide ranking of government take are not purposes in the existing laws guiding Federal oil and gas programs.

See comment 6.

The MMS analyzes fiscal terms before each lease sale and reviews the results of each sale. The analysis includes the level and extent of bidding activity to determine if fiscal terms are operating as intended. Studies and literature are reviewed related to fiscal terms and potential economic, fiscal, and geologic outcomes in different countries and regions operating under different fiscal terms. The two recent Gulf of Mexico royalty rate increases were made by the MMS incrementally in response to expected oil and gas prices. These royalty rate increases are consistent with the statutory objective to achieve fair market value. Additionally, both the MMS and the Bureau of Land Management are investigating sliding royalty options for future oil and gas leasing. The BLM is focusing on a variable royalty structure for onshore oil shale and the MMS is analyzing additional variable royalty and rental structures for the Federal OCS.

See comment 7.

The MMS recently contracted with a panel of academic oil and gas industry experts to conduct a 2-year study on issues that, to a large extent, overlap those being recommended by the GAO for a comprehensive study and independent panel. Entitled "Policies to Affect the Pace of Leasing and Revenues in the Gulf of Mexico," the study is analyzing a variety of fiscal arrangements, including fixed and sliding royalty terms much like those discussed or implied in the GAO report. A considerable part of this study is designed to cover important fiscal issues now being raised by the GAO. Interior believes it is premature and duplicative to conduct another comprehensive study of fiscal terms and convene another panel of experts until the recommendations of this panel are reviewed and evaluated.

See comment 8.

We agree that total revenue allocated to the Government (Government take) varies greatly among countries and resource owners. In many of the fiscal systems that rank higher in Government take in the studies cited in the GAO report, governments own and produce the resource. It is important to point out that global companies have choices where they make capital investments. Their investment choices are affected by many variables, including the fiscal system of rents, royalties, and bonus bids, as well as the cost of capital, risk, and the attractiveness of alternative investments. An increase in Government take through higher royalties, taxes, or other aspects of the fiscal system may result in less domestic oil and gas production than would occur at more favorable economic terms. Thus any increases in Federal revenues through higher fiscal terms must be carefully weighed. The public receives significant direct, indirect, and induced economic benefits from domestic oil and gas leasing and development in the form of jobs and economic growth, as well as less dependency on foreign energy sources.

See comment 9.

We disagree with the GAO assertion that Interior cannot properly and effectively conduct the mineral leasing programs without explicitly ranking Federal mineral leasing systems and assessing industry rates of return compared to leasing systems employed worldwide. While Government take rankings may be a comparison measure, Interior operates under a management and leasing policy defined by Congress in the OCSLA and FLPMA. Through the evaluation of varied fiscal terms, the MMS has made changes within the authority granted to the Secretary under the OCSLA and in response to subsequent legislative requirements. The OCSLA in particular describes the purposes of the offshore oil and gas program, which include expeditious exploration and development, encouraging competition, receipt of fair market value, etc.

See comment 10.

Interior notes that the report does not mention the deep water royalty relief mandated by Congress in the Energy Policy Act of 2005. The Administration has actively sought repeal of this provision. Since the GAO's findings for leasing in the OCS focus almost entirely on deep water leases in the Gulf of Mexico, the impact of the law should have been taken into consideration.

Conclusion

See comment 11.

The Department of the Interior does not fully concur with the two recommendations made to the Secretary of the Interior. While we agree that it is important to review the Department's oil and gas lease terms, Interior's existing procedures are adequate to evaluate alternative fiscal terms and systems in the context of the stated purposes, goals, and objectives of the statutory requirements. Many of the fiscal systems used by other countries are not allowed in the United States.

See comment 12.

The first recommendation, with which we do not fully concur, is that the Secretary of the Interior direct the MMS, the BLM, and other relevant agencies within Interior to conduct a comprehensive review of the Federal oil and gas fiscal system using an independent panel that includes private sector experts and oil and gas industry representatives to collect and evaluate the necessary information to make informed conclusions. Recently the MMS commissioned an outside, independent group of energy industry experts from the University of Rhode Island to conduct a related 2-year study entitled, "Policies to Affect the Pace of Leasing and Revenues in the Gulf of Mexico." Accordingly, it would be premature and duplicative for the MMS to undertake another comprehensive study, and convene another group of experts prior to completion of the current study. On any study, the Secretary will involve partners, as appropriate. The MMS and the BLM will continue to evaluate program requirements and analyze options for determining the most appropriate mineral leasing fiscal systems subject to the authorizing legislation.

See comment 13.

The second recommendation, with which we do not fully concur, is for the Secretary of the Interior to direct the MMS, the BLM, and other relevant agencies within Interior to establish procedures to periodically evaluate the Federal oil and gas fiscal system in relation to those of other resource owners and report the findings to Congress. Interior would respond to any Congressional request. The MMS and the BLM have conducted internal control reviews of key components of their fluid minerals management programs. The MMS evaluates the fiscal terms before each OCS lease sale and will continue to balance the requirements of the OCSLA and continues to respond to emerging market conditions. Interior agencies that lease Federal minerals already keep abreast of current literature on fiscal systems of other resource owners. Therefore, we believe that what other resource owners may do is useful to know, but what they do is not the most important factor to consider in designing appropriate fiscal terms for mineral leases issued by the Department of the Interior.

Through our leasing program, we continually update, evaluate, and review the fiscal terms of leases. We are always open to suggestions to clarify procedures in order to assure their effectiveness. We will closely examine those procedures internally.

Technical comments are enclosed. If you have any questions, please contact Andrea Nygren, MMS Audit Liaison Officer, at (202) 208-4343.

Sincerely,

C. Stephen Allred
Assistant Secretary

Enclosure

The following are GAO's comments on the Department of the Interior's letter dated August 8, 2008.

GAO Comments

1. Regarding Interior's statements that (1) the draft report relies heavily on measures of government take but does not clarify the link between government take and investment attractiveness, and (2) the draft report does not relate the significance of the OCSLA and FLPMA laws, we disagree. The report on page 1 states that several factors need to be considered, including the size of availability of the oil and gas resources in place; the cost of finding and developing these resources, and the stability of other the oil and gas fiscal systems and the country in general. Also on page 2, we note that a fair government take would strike a balance between encouraging private companies to invest in the development of oil and as resources on federal lands and waters while maintaining the public's interest in collecting the appropriate level of revenues from the sale of the public's resources. Further, we devote a significant portion of our discussion of objective one to how the attractiveness of the U.S. oil and gas fiscal system compares with those of other resource owners, and concludes that U.S Gulf of Mexico and other U.S. places are attractive places to invest. With regard to the significance of the OCSLA and FLPMA laws, on page 3 of the report we discuss the provisions of the OCSLA and FLPMA laws and how they relate to the management of the federal oil and gas fiscal system.
2. Interior commented that the report's conclusion that inflexibility in the federal oil and gas fiscal system is responsible for significant reductions in the federal fiscal take is not supported. We maintain that the inherent inflexibility of the federal fiscal system means that government receipts from the production of oil and gas on federal lands and waters have not tracked with the prices of oil and gas. This lack of flexibility explains, in part, why the Congress enacted the Deep Water Royalty Relief Act in 1995, a time when oil and gas prices were much lower than they are today. The lack of flexibility of the royalty rates for some of the leases issued under this Act, as implemented by Interior, will end up costing the public billions of dollars in foregone revenues. Further, the recent increases to royalty rates that Interior references in its comments do nothing to address the bulk of leases already held and for which industry profits have increased as high as they have precisely because neither the royalty rates, nor other components of the oil and gas fiscal system were sufficiently flexible to allow federal revenues to increase automatically when oil and gas prices and industry profits increased. Overall, Interior should strive to achieve fair market value over time, not simply evaluate market conditions at the time leases are issued.
3. With regard to Interior's comments that although it has not conducted a comprehensive evaluation of the federal oil and gas fiscal system, it has evaluated expected resources and conditions on the Outer Continental Shelf (offshore) tracts, we agree that Interior takes some steps to evaluate offshore leases but the objective

addresses a broader evaluation of how Interior monitors the performance and appropriateness of the entire federal oil and gas fiscal system, including offshore and onshore, and also including assessing performance over time rather than at the time a lease is sold. Interior officials told us they evaluate offshore tracts before the issuance of a lease for prospectivity of the lease and use such measures to determine an adequate minimum bid for the lease. However, as Interior makes note in its own comments, Interior officials have not systematically reviewed the bid outcomes of offshore tracts.

4. We agree that onshore and offshore leases can be very different and for the reasons stated in Interior's comments. That is why we recommended a comprehensive review of the entire federal oil and gas fiscal system, including onshore and offshore. We also recommended that the results of this comprehensive review be presented to the Congress so that it can act appropriately in the event any existing laws or regulations that govern the leasing and collection of revenues from federal oil and gas leases could be improved in light of the recommendations of the independent panel.
5. Interior states that the report implies that Interior maximizes government receipts from oil and gas leases. We disagree that the report implies this and can find no place in the report where we believe a reader would make such an inference.
6. We disagree with Interior's statement that it "analyzes fiscal terms before each lease sale and reviews the results of each sale." Interior's analysis of prospective leases is for offshore leases only and, according to Interior officials, is an analysis of the prospectivity of the offshore tract, designed to set minimum adequate bids. It is not a review of "fiscal terms," as Interior states in its comments. With regard to the two recent royalty rate increases for future oil and gas leases in the Gulf of Mexico, these increases do not resolve fair market value for past leases issued with inflexible fiscal terms and are themselves inflexible. Therefore, if future oil and gas prices turn out to be different than what Interior expected when they made the changes, the resulting outcome will again not reflect a fair return and could be too high or too low, depending on what happens in the oil and gas markets.
7. With regard to Interior's comment that it recently contracted with a panel of academic oil and gas industry experts to conduct a study of fiscal arrangements including fixed and sliding royalty terms, please see our general response to Interior's comments on page 24.
8. We agree with the statements Interior makes in this paragraph, and note that these concepts are also well represented in our report. For example, we note that investment choices are affected by many variables, including the fiscal system of rents, royalties, and bonus bids, as well as the cost of capital, risk, and the attractiveness of investments; indeed, we designed the job to discuss the first range of issues in our first objective, and the second range of issues in the second objective. We conclude, and Interior agrees, that any increases in federal revenues through higher fiscal terms must be carefully weighed; however, Interior has not done this "careful weighing" in making its royalty rate increases. That is why we recommended a comprehensive review of the federal oil and gas fiscal system.
9. With regard to Interior's comment that it operates under a management and leasing policy defined by the Congress in the OCSLA and FLPMA, we agree and this is reflected on page 3 of the draft report. However, Interior cannot effectively conduct

the mineral leasing programs without evaluating federal mineral leasing systems and assessing industry rates of return and other factors discussed in this report. Interior must keep abreast of these issues and developments in fiscal regimes elsewhere, and advise Congress on developments in the competitiveness of the federal oil and gas fiscal system versus those employed by other resource owners. Further, our audit work shows that Interior has responded to oil and gas market changes in a reactive, rather than strategic and forward-looking manner, and we believe the Congress needs to be kept abreast of changes affecting federal oil and gas leasing and revenue generation.

10. Interior comments that the draft report does not mention the royalty relief mandated by the Congress in the Energy Policy Act of 2005, and its decision to seek repeal of this provision, and that the impact of the law should have been taken into consideration. We agree that the draft report did not discuss the 2005 law explicitly but note that the results we report do implicitly take this law into consideration. Our results on the government take and attractiveness of investment in the deep water Gulf of Mexico derive largely from a 2007 study done by Wood Mackenzie that took into account the impact of the existing laws at the time of the study. We have added language to make explicit acknowledgement of the Energy Policy Act of 2005
11. See our general response to Interior's comments on page 24-25 of this report.
12. See our general response to Interior's comments on page 24-25 of this report.
13. See our general response to Interior's comments on page 24-25 of this report.

End Notes

[1] GAO, *Oil and Gas Royalties: A Comparison of the Share of Revenue Received from Oil and Gas Production by the Federal Government and Other Resource Owners*, GAO-07-676R (Washington, D.C.: May 1, 2007).

[2] In general, each country has at least one oil and gas fiscal system. Certain countries—for example, Canada and the United States—have a number of different oil and gas fiscal systems: a federal system that governs resource development on federal lands and other systems that govern resource development on provincial lands in Canada and state lands in the United States.

[3] The estimated additional revenues are the estimated reduction in the companies' share of remaining value of existing assets, when comparing fiscal systems in place at the start of 2002 and those in place in mid-2007, under a high-price scenario of $75 per barrel of crude oil.

[4] In the Wood Mackenzie study, "Gulf of Mexico" results refer only to deep water areas of 400 meters or greater depth; Wood Mackenzie currently does not have a comparable database for shallower Gulf waters.

[5] 43 U.S.C. § 1337.

[6] With regard to onshore leasing, there are both competitive and noncompetitive leases. For competitive leases, 30 U.S.C. Section 226 stipulates that a national minimum acceptable bid of $2 an acre be met and a royalty payment of not less than 12.5 percent be met, although Section 209 of the law allows the Secretary to waive or reduce rental rates or minimum royalty rates when he deems this is necessary to promote development or if the leases cannot be successfully operated under the terms provided; Section 226 of the law allows the Secretary to increase the $2 an acre minimum bid, so long as he notifies the House and Senate Committees on Natural Resources 90 days before doing so. For noncompetitive leases, if there are no bonus bids made at an auction, or if all bids are less than the national minimum, the land is offered noncompetitively, with some exceptions.

[7] Wood Mackenzie's evaluation of risk did not compare the likely resource risk from future drilling activities or include a risk comparison of technical and/or resource risks; it evaluated the risk to companies for conducting business under the specific fiscal system being evaluated.

[8] Energy Information Administration, *Performance Profiles of Major Energy Producers: 2006*, December 2007. A return of equity is another measure of company and industry profitability. A return on equity is net income divided by shareholders' equity.

[9] American Petroleum Institute, *America's Oil and Gas Industry: Putting Earnings into Perspective*, 2007.

[10] Return on investment is calculated by dividing net income by net investment in place.

[11] Our analysis differs from the other studies cited in this report because we examined return on investment for exploration and production only, instead of oil and gas industry-wide return on investment or return on equity. Additionally, our analysis of the manufacturing industry includes the universe of companies identified as manufacturers by Standard Industrial Classification code (excluding oil and gas companies) instead of an industry index as was used by the American Petroleum Institute and EIA studies. As a result, our return on investment differs from the other studies.

[12] World Bank, *Doing Business 2008: Comparing Regulation in 178 Economies*, Washington, D.C., 2007.

[13] Six of the 25 companies that have received royalty relief to date have signed agreements with Interior to allow the inclusion of price thresholds for leases signed in 1998 and 1999. A list of U.S. and international companies that currently receive royalty relief, and who may be affected by the outcome of the legal challenge, is presented in app. II.

[14] GAO, *Oil and Gas Royalties: Litigation over Royalty Relief Could Cost the Federal Government Billions of Dollars,* GAO-08-792R (Washington, D.C., June 5, 2008).

[15] By foregone revenue, we mean the royalty revenue that would have accrued to the federal government had there been no royalty relief under the DWRRA.

In: Royalties at Risk in the Development...
Editor: Thomas Marinelli

ISBN: 978-1-60692-837-0

Chapter 2

MINERAL REVENUES: DATA MANAGEMENT PROBLEMS AND RELIANCE ON SELF-REPORTED DATA FOR COMPLIANCE EFFORTS PUT MMS ROYALTY COLLECTIONS AT RISK*

U. S. Government Accountability Office

September 12, 2008

Congressional Requesters

Subject: *Mineral Revenues: Data Management Problems and Reliance on Self-Reported Data for Compliance Efforts Put MMS Royalty Collections at Risk*

The Department of the Interior's (Interior) Minerals Management Service (MMS) collected the equivalent of over $9 billion in oil and gas royalties in fiscal year 2007, more than $5 billion of which it deposited in the U.S. Treasury; it dispersed the remaining approximately $4 billion to other federal, state, and tribal accounts. These royalties—payments made to the federal government for the right to produce oil and gas from federal lands and waters—represent one of the country's largest nontax sources of revenue. The amount of oil and gas royalties MMS collects may increase if the price of energy increases and industry's demand to drill on lands and in waters controlled by the federal government continues to trend upward. For example, the price of West Texas Intermediate—a type of oil commonly used as a benchmark—has risen by more than 100 percent since January 2007 and recently exceeded $140 per barrel, a price that, when adjusted for inflation, is at the highest level since 1980. Moreover, applications for onshore drilling permits at Interior's Bureau of Land Management (BLM) continue to increase, from approximately 4,000 in 2001 to more

* This is an edited, reformatted and augmented version of U. S. Government Accountability Office publication, 97-471, dated September 12, 2008.

than 10,000 in 2007. Similarly, offshore leasing activity overseen by MMS's Offshore Energy and Minerals Management (OEMM) has also generally increased since 2001.

Companies that develop and produce oil and gas resources from federal lands and waters do so under leases obtained from and administered by Interior—BLM for onshore leases and MMS's OEMM for offshore leases. Together, BLM and OEMM are responsible for ongoing oversight of oil and gas operations on more than 28,000 producing leases to help ensure that oil and gas companies comply with applicable laws, regulations, and agency policies. Among other things, BLM and OEMM staff inspect leases to verify that oil and gas production is accounted for as required by the Federal Oil and Gas Royalty Management Act of 1982[1] and agency regulations and policies. These inspections typically include an examination of the meters and their calibration records. Additionally, BLM inspectors may review production records, forwarding any discrepancies to MMS for resolution. OEMM has a slightly different process; after reviewing all production records, its production verification team addresses any gas volume discrepancies itself, forwarding oil volume discrepancies to MMS. The results of these inspections and reviews are recorded in the agencies' management databases.[2]

As a condition of producing oil and gas under federal and Indian leases, companies are required to file two key monthly reports with MMS—one specifying the total production and disposition of oil and gas and the other stating the royalties due on that production. However, because of various leasing and development arrangements made by companies, these two reports are often submitted by different companies. The companies physically developing the lease, or the operators, are responsible for reporting the production volumes to MMS in their monthly production reports.[3] The companies with a financial interest in that lease, or the payors, are responsible for reporting the cash royalty owed on the federal and Indian oil and gas production in their monthly royalty reports.[4] Each month, payors calculate the royalty payment they owe to the federal government using the four key variables illustrated in the following equation:

Royalty payment = (sales volume x sales price - deductions) x the royalty rate[5]

Companies enter the monthly production and royalty reports via a Web-based portal to MMS's royalty information technology (IT) system and may make adjustments to those entries for up to 6 years after the initial reporting date.[6] In addition to filing the royalty report with MMS, payors typically make the actual cash royalty payment via an electronic fund transfer to an account at the Department of the Treasury (Treasury). Once MMS reconciles the self-reported royalty payment data from the monthly royalty reports with the payments submitted to Treasury, MMS disburses the royalties from the Treasury account to the appropriate federal, state, tribal, and allotted accounts.[7] All of these transactions are recorded and stored in MMS's current royalty IT system, which went on line in 2001. Since then, MMS's IT system has experienced several problems but has continued to improve. For example, MMS has worked for the past several years to enhance the royalty IT system by both refining and increasing the number of edit checks designed to prevent companies from entering incorrect royalty and production data.

As an additional check on the accuracy of both the company-reported production and royalty data, MMS conducts either audits or compliance reviews on these data within 3 years of their submission. For example, during fiscal year 2008, MMS is conducting compliance work on calendar year 2005 payments. Audits are an assessment of the accuracy and completeness of the self-reported production and royalty data compared against third-party documents, such as sales contracts and oil and gas sales receipts from pipeline companies. By contrast, compliance reviews assess the data's reasonableness—a quicker, more limited check of the accuracy and completeness of a company's self-reported data—and they do not include a systematic examination of underlying third-party documentation. In addition, some states and tribes that receive royalties collected by MMS have agreements with MMS authorizing them to conduct either audits or compliance reviews on federal and Indian producing leases within their jurisdictions.[8] Under current law, MMS has 7 years in which to make monetary demands on federal royalty payors.[9] Historically, MMS has annually assessed its overall compliance performance on the basis of whether it has conducted compliance activities—either full audits or compliance reviews—on leases or companies that are responsible for generating a predetermined percentage of royalty payments within 3 years after the royalty payment is due. This approach resulted in MMS auditing many of the same high royalty paying companies and leases year after year, while conducting only limited compliance work on other companies during that same time period. To ensure greater compliance coverage on both companies and leases, MMS is now in the process of implementing a new risk-based compliance approach that will assist in selecting companies to audit or review based on factors in addition to royalties paid.

Given the financial importance of royalty management, MMS has been the subject of considerable scrutiny through the years by entities such as GAO; Interior's Inspector General (IG); and the Royalty Policy Committee (RPC), a group empanelled by the Secretary of the Interior and charged with providing advice on managing federal and Indian leases and revenues. For example, the RPC issued a report in December 2007 that included more than 100 recommendations to strengthen Interior's royalty collections by improving BLM's and OEMM's production accountability practices, MMS's compliance efforts, and MMS's and BLM's interagency coordination. As part of Interior's response to these recommendations, it convened staff from the relevant agencies and developed detailed plans for addressing these recommendations, some of which it has already implemented. In addition, in March 2008, we testified before the House Committee on Natural Resources, Subcommittee on Energy and Mineral Resources, on our preliminary findings on Interior's royalty collection program. This report follows up on our prior work and includes recommendations to strengthen Interior's royalty collections program. Specifically, you asked us determine (1) whether Interior has adequate assurance that federal oil and gas are measured accurately, (2) whether MMS's royalty IT system and royalty collection and verification processes provide sufficient assurance that all royalties are being collected, and (3) the extent to which MMS's compliance efforts provide an adequate check on industry's self-reported data.

To address Interior's oil and gas production measurement accountability practices, we reviewed OEMM's and BLM's documentation on policies and procedures for conducting production inspections and production verification work. In addition, we interviewed MMS inspectors in the Gulf of Mexico, Pacific, and Alaska regional offices and toured oil and gas production facilities in the Gulf of Mexico. We further interviewed BLM state officials in five states with significant federal oil and gas production and interviewed BLM officials from five BLM field offices selected from a nonprobability sample of the 27 BLM field offices overseeing oil and gas production in those five states. The BLM field offices were selected based on geographic location, the number of violations of regulations, and the number of oil and gas volume errors identified during production inspections. We also toured oil and gas production facilities in Wyoming and Colorado. To assess MMS's royalty IT system and royalty collection and verification processes, we reviewed MMS's documentation and processes for collecting royalty and production data and examined how that information is stored and manipulated in MMS's royalty IT system. We also interviewed MMS staff regarding these processes. Finally, to assess MMS's compliance efforts, we reviewed MMS's audit and compliance manuals and its newly revised procedures, interviewed compliance staff, and attended several demonstrations of MMS's compliance IT system. We also sent questionnaires addressing production, royalty data, and compliance issues to the 11 state and 7 tribal members of the State and Tribal Royalty Audit Committee; 9 states and 5 tribes responded. We conducted this work from April 2007 to July 2008 in accordance with generally accepted government auditing standards. Those standards require that we plan and perform the audit to obtain sufficient, appropriate evidence to provide a reasonable basis for our findings and conclusions based on our audit objectives. We believe that the evidence obtained provides a reasonable basis for our findings and conclusions based on our audit objectives.

Summary

Neither BLM nor OEMM is meeting statutory obligations or agency targets for conducting inspections of certain leases and metering equipment used to measure oil and gas production, raising uncertainty about the accuracy of oil and gas measurement. Moreover, when these inspections have been conducted, BLM and OEMM have at times recorded inspections inaccurately in their databases. Specifically, although BLM and OEMM are statutorily required to annually inspect leases producing "significant quantities of oil or gas" and those with a "history of noncompliance," the Secretary has defined these terms for onshore leases but not for offshore leases. Understanding the meaning of these terms is necessary to implement the act and is critical to the agencies' ability to prioritize inspections. Although BLM is able to prioritize its inspections, according to BLM officials, they are not completing all of the inspections required by law and agency policy, in part because their workload has substantially grown because of increased onshore drilling. OEMM, on the other hand, is not able to prioritize its inspections because the statutory terms have yet to be defined by the Secretary. Moreover, OEMM is not meeting its agency targets for inspections because, according to OEMM officials, inspectors are still conducting cleanup activities in the Gulf of Mexico—where almost all of the offshore oil and gas production occurs—in the wake of Hurricanes Katrina and Rita in 2005. Finally, although it is important that data are recorded accurately, when both BLM and OEMM conduct production inspections, neither agency is consistently doing so. Officials from both BLM and OEMM told us that some inspection and data entry staff are relatively inexperienced and do not always record the inspections in their databases as procedures require. Accurate data are necessary not only to monitor progress throughout the year to determine whether annual goals are achieved, but also to assist MMS in its royalty compliance activities. We are making several recommendations to the Secretary of the Interior to provide greater certainty that oil and gas produced from federal lands and waters are measured accurately.

MMS's royalty IT system and processes for collecting and verifying royalty data have improved, but they continue to lack several capabilities that would provide greater assurance that royalties are being accurately collected. For example, MMS's royalty IT system cannot monitor adjustments made to production and royalty data by companies. While MMS is working to address this issue, companies may continue to adjust their previously self-reported production and royalty data without prior MMS approval or review. This includes adjustments made by companies to data after MMS completes its compliance work, meaning that while the royalties paid were accurate at the close of the audit, they may not remain accurate. Furthermore, MMS is unable to identify, in a timely manner, instances in which a royalty report has not been submitted by a company, and, as a result, MMS cannot be entirely confident it is receiving all of the royalties when they are due. Finally, MMS lacks a clear process to determine that royalties are accurately paid in instances when OEMM or BLM identify volume discrepancies during their production inspections and verification work. For example, when BLM identifies an over- or under-reporting of production volumes, BLM notifies the production reporting section of MMS. While MMS staff may work to correct the production numbers, staff do not relay this information to the royalty reporting section so that staff can check that the appropriate royalties were paid. To provide greater assurance that MMS is accurately collecting royalties in a timely manner, we are making several recommendations aimed at improving its royalty IT system and royalty collection and verification processes.

While MMS continues to strengthen its compliance efforts, MMS's use of compliance reviews, which are more limited in scope than audits, has led to an inconsistent use of third-party documents to verify that self-reported industry production and payment data are correct, thereby placing royalty collections at risk. MMS has historically relied on audits to determine whether a company accurately paid its royalties by examining third-party documents that contained information on prices, volumes, and deductions. More recently, MMS has transitioned to relying heavily on compliance reviews that assess whether the royalties paid by a company are reasonable, and do not always include an examination of third-party documents. Furthermore, while MMS's compliance reviews of offshore leases include a systematic comparison between a company's reported production volumes and independent pipeline company documents, an analogous process does not exist for onshore leases, in part because of the significantly greater numbers of leases and pipelines for which data would have to be collected. The absence of a consistent check on self-reported data—such as comparing the data with third-party documents—when conducting onshore compliance reviews raises questions about the accuracy of royalty payments. To address this issue, we are recommending that MMS require that third-party documents be reviewed when conducting onshore compliance reviews.

BLM and OEMM Are not Completing the Required Production Inspections, Leaving the Accuracy of Oil and Gas Measurements in Doubt

Interior lacks adequate assurance that federal oil and gas volumes are being measured accurately because neither BLM nor OEMM is fully inspecting leases and metering equipment as required by law and agency policies; the agencies also are not always entering accurate inspection data into their databases. BLM is charged with inspecting approximately

20,000 producing onshore leases annually to ensure that oil and gas volumes are accurately measured. However, BLM's state inspection and enforcement coordinators from Colorado, Montana, New Mexico, Utah, and Wyoming, which are responsible for more than 95,000 wells, reported that only 8 of the 24 field offices in the five states completed both their (1) required annual inspections of leases that are high-producing and those that have a history of violations—terms defined by Interior's policies for onshore leases so that BLM can implement the Federal Oil and Gas Royalty Management Act and (2) inspections every third year on all remaining leases as required by BLM policy.[10] According to the BLM state inspection and enforcement coordinators, the number of completed production inspections varied greatly by field office. For example, while BLM inspectors were able to complete all of the production inspections in the Kemmerer, Wyoming, field office, inspectors in the Glenwood Springs, Colorado, field office were able to complete only about one-quarter of the required inspections. Officials in three of the five field offices in which we held detailed discussions with inspection staff told us that they had not been able to complete the production inspections because of competing priorities.[11] For example, BLM officials told us activities related to drilling, including drilling inspections, take priority over completing production inspections. Further, with the significant number of wells being drilled, drilling-related activities demand much of the inspectors' time. BLM officials also discussed the high turnover rate for inspection staff caused by the high-pressure work environment, the ability to make more money in the private sector, and the high cost of living in many areas where BLM has field offices. Importantly, BLM officials from all five field offices told us that when they have conducted production inspections, they have identified numerous violations. For example, BLM staff in four of the five field offices identified errors in the oil and gas production volumes reported by operators to MMS by comparing production reports with third-party source documents. Additionally, BLM staff from one field office we visited showed us a bypass built around a gas meter that could allow gas to flow around the meter without being measured. BLM staff subsequently ordered the operator to remove the bypass but did not fine the operator. Staff from another field office told us of a case in which BLM had to remove illegal gas lines that routed gas to private residences. Finally, in one of the field offices we visited, BLM officials told us of an instance in which a company maintained two sets of conflicting production data—one used by the company and another reported to MMS. In this instance, BLM ordered the company to correct and resubmit production data to MMS.

Similarly, OEMM, which is responsible for inspecting offshore production facilities that include oil and gas meters, did not complete inspections of offshore oil and gas royalty meters as required by its policy or applicable law in 2007. OEMM officials responsible for inspecting royalty meters in the Gulf of Mexico—meters that measure oil and gas for determining royalty payments—told us they completed about half of the required 2,700 royalty meter inspections in 2007. This is far less than required under OEMM's 1994 Gulf of Mexico production inspection policy, which requires all oil and gas royalty meters be inspected annually. Meter inspections are an important aspect of the offshore production verification process because, according to OEMM officials, one of the most common violations identified during inspections is missing or broken meter seals. These seals are required to be kept in place to prevent tampering with measurement equipment. When seals are broken, without closer inspection and testing, it is not possible to determine whether the meter is correctly measuring oil or gas production. OEMM officials explained that one reason

OEMM failed to complete its required inspections was the continuing cleanup work related to Hurricanes Katrina and Rita in 2005. Furthermore, the Federal Oil and Gas Royalty Management Act instructed the Secretary of the Interior to inspect leases annually that are producing or expected to produce significant quantities of oil or gas in any year or that have a history of noncompliance with applicable law or regulations. While Interior has defined these terms for onshore oil and gas leases, it has not yet done so for offshore oil and gas leases. As a result, when production inspectors are unable to inspect all royalty meters as required by their policy, they lack the criteria for effectively prioritizing inspections based on production volumes or concerns about noncompliance. Therefore, the Secretary cannot be assured that federal oil and gas production is being measured accurately.

Moreover, when BLM and OEMM do complete production inspections, staff are not always accurately recording this work in their databases. For example, of the 27 BLM field offices in five states we reviewed, 2 field offices had fiscal year 2007 production inspection data that could not be validated by the state's inspection and enforcement coordinator, and 6 field offices had instances of production inspection data changing after the end of the fiscal year, in violation of BLM's policy. For example, Wyoming's state inspection and enforcement coordinator could not validate a field office's inspection numbers because the coordinator lacked confidence in that field office's ability to correctly identify high production leases. The coordinator explained that because of MMS and BLM data transfer issues caused by ongoing Indian trust fund litigation, BLM staff may not be able to easily or accurately identify high producing leases.[12] Additionally, 6 field offices revised their completed production inspection data after the end of the fiscal year. For example, officials in a Montana field office reported that 20 additional inspections were entered into the database for fiscal year 2007 because staff identified inspections that had been completed and documented in paper files but had not been entered into their database. Finally, in a number of instances, the inspection priority changed during the year. For example, one of the field offices in Montana adjusted the number of high-priority inspections downward after staff realized they had initially incorrectly identified the high- priority inspections because of database shutdowns caused, again, by the ongoing Indian trust fund litigation.

Similarly, according to interviews with OEMM officials, OEMM staff do not consistently check the accuracy of the production inspection data entered. For example, OEMM data indicated that at one facility no meters were available for inspection, yet 16 meter inspections had occurred during calendar year 2007, which an OEMM official confirmed was an error. An OEMM official explained that one reason for errors may be data entry problems (a clerk, not the inspectors themselves, enter the data). OEMM officials state they are currently implementing new procedures to ensure that valid data are entered into the database. Accurate and timely data are critical for monitoring ongoing progress and whether annual goals and targets are met. However, because of the errors identified in both BLM's and OEMM's data, we cannot report the precise number of production inspections completed.

MMS's Royalty Information Technology System and Processes are Improving But Still Lack Key Functions That Could Provide Greater Assurance on the Accuracy of Royalty Collections

While MMS continues to improve its current royalty IT system, which went online in 2001, and its royalty collection processes, it still lacks several key functions and processes that could provide greater assurance that royalties are accurately collected, including, (1) the ability to maintain the accuracy of production and royalty data entered by companies, (2) the ability to identify missing royalty reports in a timely manner, and (3) a process to ensure that accurate royalties are paid on volume discrepancies identified by OEMM and BLM during their production inspections.

MMS's ability to maintain the accuracy of production and royalty data is hampered because companies can make adjustments to their previously entered data without prior approval. Companies may legally make changes to both royalty and production data in MMS's royalty IT system for up to 6 years after the initial reporting month, and these changes may necessitate changes in the royalty payment.[13] According to MMS, most adjustments are made within the first several years after initial reporting. For example, MMS's analysis of adjustments made to royalty reports in fiscal year 2007 show that 75 percent of all accepted adjustment lines were reported within 3 years of their corresponding sales month. However, MMS's royalty IT system currently allows companies to make adjustments to their data beyond the allowed 6-year time frame. MMS examined data from September 2002 through July 2007 to determine the number of adjustments made outside the 6-year time frame. The results of this analysis suggested that more than 81,000 adjustments were made to data outside the time frame, though according to MMS officials, some of these adjustments were retroactively approved. As a result of the companies' ability to make these retroactive changes, within or outside of the 6-year time frame, the production data and required royalty payments can change over time, complicating efforts by agency officials to reconcile production data and ensure that the proper royalties were paid. Furthermore, changes made to the data do not necessarily trigger a review to determine their reasonableness or whether the adjustments impact the royalties due. According to agency officials, these adjustments are investigated by staff only if the company or lease is selected for an audit or compliance review. This is problematic for several reasons. First, current law is that companies may make adjustments to their federal royalty payment data up to 6 years after the production reporting month.[14] MMS issues monetary demands for unpaid royalties up to 7 years after the last adjustment made by a company. However, the law, which provides that a demand be made within 7 years from the date an obligation becomes due, could be interpreted as giving MMS only 7 years from the initial month of the production reporting date.[15] This interpretation would, in some cases, allow MMS only 1 year to identify any erroneous adjustments and issue a monetary demand for any additional royalties due. Second, companies may change production and royalty data after an audit or compliance review has been completed, making it unclear whether these audited royalty payments remain accurate after they have been reviewed. MMS typically conducts compliance work approximately 3 years after companies initially report the data, at which point, according to MMS officials, companies have made the majority of their adjustments. However, companies have several additional years after MMS typically completes its compliance work to adjust their royalty data, and current MMS policies do not require that all subsequent postcompliance work adjustments be reviewed. MMS is aware of these issues and is in the process of addressing the issues over which it has authority. Specifically, MMS is developing requirements to modify its royalty IT system to monitor adjustments. The stated goal is for the royalty IT system to automatically identify adjustments that have been made to data outside of the allowable 6-year time frame and to monitor adjustments made within the allowable time frame after MMS closes an audit or compliance review.

Further, MMS's royalty IT system is unable to automatically detect instances when a royalty payor fails to submit the required royalty report in a timely manner. MMS relies on two critical pieces of data in verifying the accuracy of royalty payments: the production report, which operators submit to document the total oil and gas production from a particular lease or several combined leases, and the royalty report, payors submit to document the royalty paid on oil and gas production. With few exceptions, MMS should receive a corresponding royalty report or reports for each production report it receives. When MMS's current royalty IT system went online in 2001, it was not able to reliably detect either missing production or royalty data. In 2004, MMS modified its royalty IT system to automatically detect missing production data. As a result, MMS has identified a backlog of approximately 300,000 missing production data records, which includes both entire production reports and missing wells on those production reports. MMS subsequently established teams to research the missing records and told us that the goal was to resolve all missing record issues older than 90 days by the end of 2008. More recently, however, MMS has revised its goal downward to address 80 percent of the missing records by the end of 2008. Preliminary results, according to agency staff, suggest that many of the issues can be explained by BLM staff delays in entering well and production reference data and sharing that data with MMS, and not necessarily by a company's failure to submit the required report. However, MMS's royalty IT system continues to lack the ability to automatically detect cases in which an expected royalty report has not been filed in a timely manner. As a result, cases in which a company stops filing royalty reports and stops paying royalties may not be detected until more than two years after the initial reporting date, when MMS's royalty IT system completes a reconciliation of volumes reported on the production reports with the volumes on their associated royalty reports. Therefore, it remains possible under MMS's current strategy that the royalty IT system may not identify instances in which a payor stops reporting until several years after the report is due. This creates an unnecessary risk that MMS may not be collecting accurate royalties in a timely manner.

Finally, MMS lacks a consistent process to ensure that the appropriate royalties are paid when either OEMM or BLM identifies volume discrepancies while conducting production inspections and other production verification activities. To verify that oil and gas production is accurately reported to MMS, BLM sometimes and OEMM always compares the volumes reported on the production report to other third-party documents that contain production information, such as a pipeline statement. When they find discrepancies between these two documents, they typically forward the information to MMS, which then takes steps to reconcile and correct the discrepancies. For example, BLM's policy is for staff to forward all volume discrepancies to MMS's royalty reporting section, which then attempts to correct the volume error by talking to operators. In some instances, MMS may request that the operator submit a corrected production report. However, MMS staff in the production reporting section do not automatically notify the royalty reporting staff so that they may check to ensure that the correct royalties were paid. In other words, these newly reconciled data are not automatically and systematically compared in a timely manner with the reported sales volume in the royalty report previously entered into the royalty IT system. While a comparison between these revised production data and the production data included in the royalty report may occur several years later via MMS's production and royalty report volume reconciliation process, or if the royalty payor's property has been selected for an audit or compliance review, without a timely systematic comparison of all such records, MMS cannot ensure that the initial royalty payment is accurate.

MMS Does Not Consistently Review Third-Party Documents for Onshore Compliance Reviews to Verify Company-Reported Royalty and Production Data aAre Accurate

While MMS is currently strengthening its process for selecting companies or leases to review, its increased use of compliance reviews—which are more limited in scope than audits—has led to an inconsistent use of third-party documents to verify that self-reported royalty data are correct, putting accurate royalty collections at risk. According to MMS, compliance reviews can be conducted much more quickly and require fewer resources than audits, largely because they represent a more limited reasonableness check of the accuracy and completeness of a company's self-reported data and do not include a systematic examination of underlying third-party documentation. Audits, on the other hand, are more time- and resource-intensive, and they include the review of third-party documents, such as sales revenue data, transportation and gas processing costs, and production volumes, to verify whether company-reported data are accurate and complete. Finally, a 2006 Interior IG audit found that MMS's data tracking the number of completed audits and compliance reviews were inaccurate, and as such, we are unable to provide this information.

Currently, OEMM requires that companies submit pipeline production data for oil and gas produced offshore. MMS then uses the data when conducting compliance reviews for offshore properties. Additionally, MMS recently assessed the effectiveness of its offshore compliance reviews. For calendar year 2002, MMS compared the results of 100 out of about 700 compliance reviews of offshore leases and companies with the results of audits conducted on those same leases or companies. As a result of this evaluation, MMS now plans to improve its compliance review process by, for example, ensuring that it includes a step to check that royalties are paid on all royalty-bearing products, including other petroleum-based products such as retrograde—a liquid product that condenses out of natural gas.

In contrast, because of the significantly greater number of onshore leases and pipelines, BLM collects only a limited amount of pipeline and other third-party data for onshore production. Furthermore, because MMS itself does not routinely request these data from the companies, it does not systematically use third-party data when conducting onshore compliance reviews. In 2006, Interior's IG reviewed MMS's compliance process and made a number of recommendations to strengthen it. The IG recommended, among other things, that MMS examine 1 month of third-party documentation as part of each compliance review to provide greater assurance that both the production and allowance data are accurate. To address this recommendation, MMS revised its compliance review guidance to include suggested steps for reviewing third-party source production data, when available, for both offshore and onshore oil and gas. However, the revised guidance falls short of making these steps a requirement for onshore compliance reviews. And while MMS completed a study comparing the results of compliance reviews with those of audits for offshore properties and leases, MMS could not provide us with a similar study for onshore properties and leases.

Moreover, as noted above, as of 2001, MMS began using compliance reviews in addition to audits to meet its performance goal of completing compliance activities on a specified percentage of royalty payments within 3 years of the initial royalty payment. For example, in 2006, MMS reported that it had achieved its goal by confirming reasonable compliance on 72.5 percent of all calendar year 2003 royalties.[16] However, Interior's IG found in a 2006 audit that MMS did not track the extent to which it achieved its performance goal through audits that systematically rely on third-party documents, as opposed to compliance reviews that do not. Additionally, the IG found that MMS's data on completed audits and compliance reviews were inaccurate and incomplete. When we examined MMS's compliance program in 2007, MMS did not yet have reliable data on either the numbers of audits and compliance reviews completed or their respective contribution to the annual performance goal. MMS is now in the process of implementing Interior's IG recommendation and should have these data available for reporting purposes in 2009. During the same audit, Interior's IG also found that to help meet its performance goal, MMS had historically conducted compliance reviews or audits on the leases and companies that generated the most royalties, with the result that the same leases and companies were reviewed year after year. Accordingly, many leases and companies have gone years without having been audited, though some were subject to a more limited review. To address this compliance gap, Interior's IG recommended that MMS develop risk criteria for selecting leases and companies to conduct either compliance reviews or audits. MMS responded by working with a contractor to identify the necessary criteria for developing a risk-based approach for selecting companies and leases for either audits or compliance reviews. MMS is now implementing a pilot project, after which it will assess the results to further refine its new compliance approach.

Finally, representatives from the states and tribes who are responsible for conducting compliance work under agreements with MMS have expressed concerns about the quality of self-reported production and royalty data they use in their reviews. As part of our work, we sent questionnaires to all 11 states and 7 tribes that conducted compliance work for MMS in fiscal year 2007. Of the 9 state and 5 tribal representatives who responded, 7 reported that they lack confidence in the accuracy of the royalty data. For example, several representatives reported that because of concerns with MMS's production and royalty data, they routinely look to other sources of corroborating data, such as production data from state oil and gas agencies and tax agencies. Finally, several respondents noted that companies frequently report production volumes to the wrong leases and that they must then devote their limited resources to correcting these reporting problems before beginning their compliance reviews and audits.

Conclusions

Interior has been subject to significant examination and oversight of its royalty management programs over the years, which has resulted in GAO, Interior's Inspector General, and the Royalty Policy Committee issuing numerous recommendations to improve royalty collections. So far, Interior has been responsive to these recommendations and, as an example, is currently implementing an action plan to address the Royalty Policy Committee's recently issued recommendations. As a result, many of Interior's processes and systems are in flux, and the outcome of these potential improvements will not be known for some time. However, given high oil and gas prices and the increased interest on the part of oil and gas companies in the nation's oil and gas resources, it is important that we have a royalty collection system going forward that can assure the American public that the government is receiving accurate and timely royalty payments. Critical to this is that both BLM and OEMM complete and accurately document their production inspection and verification work. Furthermore, collections of accurate royalties will remain at risk as long as companies may make unverified adjustments to royalty and production data after MMS completes its compliance activities. Increasing this risk is uncertainty regarding the statutory time frames for MMS to collect unpaid royalties, which under one interpretation may leave just 1 year for MMS to identify an improper adjustment. Ultimately, Interior's royalty IT system and policies should provide adequate assurance that the federal government receives appropriate value for oil and gas produced from federal lands and waters. This royalty collection process should also rely less on companies providing accurate information on production and royalties owed, and more on a system with the ability to conduct thorough and independent verification of what is owed to the government, using third-party data where available at reasonable cost, and more systematically examining company source documentation.

Recommendations for Executive Action

To help provide greater assurance that federal oil and gas is being measured accurately, we recommend the Secretary of the Interior take the following three actions:

- Report to Congress any year in which OEMM and BLM have not met their legal and agency requirements for completing production inspections, along with the cause and a plan for achieving compliance.
- Define the terms "lease sites producing or expected to produce significant quantities of oil or gas in any year" and "lease sites which have a history of noncompliance with applicable provisions of law or regulations" for offshore oil and natural gas leases.
- Direct BLM and OEMM to evaluate both the accuracy and completeness of production inspection data in their databases, including the timeliness of data entry, and amend relevant policies and procedures as necessary.

In addition, we recommend that the Secretary of the Interior direct MMS to take the following three actions to improve its royalty IT system and royalty collection and verification processes:

- Conduct a study on the Federal Oil and Gas Royalty Simplification and Fairness Act's effect on MMS's capacity to efficiently and accurately collect federal royalties due by analyzing both the (1) 6-year timeframe for allowing companies to make adjustments to their federal royalty data and (2) MMS's 7-year time frame for issuing monetary demands for additional royalties. This study should identify an appropriate time period cutoff for allowing companies to make adjustments without MMS's prior approval to their royalty and production data and related payments, address the need for clarification on when the 7-year time period begins for issuing a monetary demand, and report the findings to Congress.
- Finalize the adjustment line monitoring specifications for modifying its royalty IT system and fully implement the IT system so that MMS can monitor adjustments made outside the legal 6-year time frame, and ensure that any adjustments made to production and royalty data after compliance work has been completed are reviewed by appropriate staff.
- Develop processes and procedures by which MMS can automatically identify when an expected royalty report has not been filed in a timely manner, and contact the company to ensure it is complying with both applicable laws and agency policies.

Finally, to improve its compliance program, we recommend that the Secretary of the Interior direct MMS to require that the onshore compliance review process include the review of a sample of third-party documentation in instances when BLM has not already collected this information to provide additional assurance that self-reported data are correct.

Agency Comments and Our Evaluation

We provided a draft of this report to the Department of the Interior for review and comment. The Department of the Interior provided written comments that are presented in Enclosure I. DOI generally agreed with our findings and conclusions. In addition, they concurred with six of the seven recommendations, and partially concurred with the other recommendation. For the recommendation on which they partially concurred regarding reporting to Congress on production inspections annually, DOI agreed with the need to report to Congress when production inspections were not completed as required, but disagreed on the nature of the reporting we recommended in our draft report. Specifically, Interior commented that it would be more useful to provide Congress with a report only in years in which they fail to meet their inspection requirements rather than providing Congress with an annual report on whether production inspections were completed along with an explanation as to the why the inspections were not completed. We modified the recommendation in response to this comment.

Frank Rusco
Acting Director, Natural Resources
and Environment

Jeanette Franzel
Director, Financial Management and
Assurance

List of Requesters

The Honorable Jeff Bingaman
Chairman
Committee on Energy and Natural Resources
United States Senate

The Honorable Nick J. Rahall, II
Chairman
Committee on Natural Resources
House of Representatives

The Honorable Tom Davis
Ranking Member
Committee on Oversight and Government Reform
House of Representatives

The Honorable Darrell Issa
Ranking Member
Subcommittee on Domestic Policy
Committee on Oversight and Government Reform
House of Representatives

The Honorable Carolyn Maloney
House of Representatives

ENCLOSURE I

United States Department of the Interior

OFFICE OF THE SECRETARY
Washington, DC 20240

AUG 2 8 2008

TAKE PRIDE IN AMERICA

Mr. Frank Rusco
Acting Director, Natural Resources and Environment
Government Accountability Office
441 G Street, NW
Washington, D.C. 20548

Dear Mr. Rusco:

Thank you for the opportunity to review and comment on the Government Accountability Office draft report entitled, "*MINERAL REVENUES: Data Management Problems and Reliance on Self-Reported Data for Compliance Efforts Put Royalty Collections at Risk*" (GAO-08-893R), which follows up on prior GAO work. We generally agree with your findings and concur with six of the seven recommendations in the draft report. We partially concur with the report's first recommendation to report annually to the Congress the extent to which we have met legal and agency requirements for completing production inspections. Reporting on exceptions rather than instituting an additional annual report as recommended by the GAO provides more relevant information to the Congress. Exception reporting (such as when a goal is missed in the event that yearly production inspection requirements are not met) highlights the exception making the report more relevant and useful. The MMS already provides annual reports to the Congress on minerals revenue management. Therefore, instead of an additional annual report to the Congress, a report will be sent to the Congress with an explanation of the circumstances contributing to missed goals. We would be happy to provide any report requested by Congress.

GENERAL COMMENTS

As noted in your draft report, the Department of the Interior's Minerals Management Service and Bureau of Land Management are responsible for the administration and oversight of oil and gas operations, including revenue collection from more than 28,000 producing Federal leases onshore and offshore. The MMS and the BLM work together to ensure that oil and gas companies comply with applicable laws and pay their correct royalty on Federal oil and gas leases.

The draft report notes that the MMS royalty information technology system and processes continue to improve. At the same time, the MMS continues to strengthen its compliance efforts. Recently, the MMS developed and deployed a risk-based analysis tool to focus its resources on auditing and reviewing companies and properties with the greatest risk of underreporting oil and gas production and the resulting royalties due to the Federal Government.

As you noted, the Department has been responsive to recommendations made by various oversight groups. For example, the draft report notes that Secretary of the Interior Kempthorne appointed the Royalty Policy Committee, Subcommittee on Royalty Management, to conduct an independent, bipartisan examination of the minerals revenue management program. The

Subcommittee, co-chaired by former Senators Bob Kerrey (D-NE) and Jake Garn (R-UT), developed a report with more than 100 recommendations to improve the Department's mineral royalty program. The Department is aggressively implementing the Subcommittee's recommendations and has already taken steps to enhance the recommendations related to production accountability and collections.

The draft report recognizes that the MMS and the BLM collect third-party and source documentation to validate the accuracy of the royalty and production data reported by Federal and Indian oil and gas lessees. However, both agencies can enhance the use and timeliness of this information to provide greater assurance that royalties are being accurately collected.

CONCLUSION

We are committed to securing America's energy future and quality of life while protecting the environment. The DOI promotes energy, the economy, and the environment. We look forward to continuing to work with you to improve the accuracy of royalty data collections, and we will provide any reports that Congress requests. If you have any questions about this response, please contact Andrea Nygren, MMS Audit Liaison Officer, at (202) 208-4343.

Sincerely,

for C. Stephen Allred
Assistant Secretary
Land and Minerals Management

End Notes

[1] Pub. L. No. 97-451, § 101 (1983).

[2] BLM records inspection information in its Automated Fluid Minerals Support System, while OEMM records its inspection information in its Technical Information Management System.

[3] Companies are required to self-report monthly production volumes to MMS on an Oil and Gas Operations Report (OGOR) form.

[4] Companies are required to self-report monthly royalty payments to MMS on the Report of Sales and Royalty Remittance Form, Form 2014.

[5] The royalty rate varies somewhat but is typically in the range of 12.5 percent to 18.75 percent. In other words, the federal government typically receives between 12.5 percent and 18.75 percent of revenue less allowable deductions, for oil and gas produced on federal lands and waters. Allowable deductions include payments to pipeline companies and other shipping costs required to transport the commodity to a market center, as well as adjustments made for the costs of processing natural gas.

[6] This system, also known as the Minerals Revenue Management Support System, is designed to store and support the collection, verification, and disbursement of royalty revenues from federal and Indian mineral leases. Indian leases do not have time restrictions on making adjustments.

[7] An allotted account is an account set up to receive royalties for any land held in trust or restricted status by the Secretary of the Interior, on behalf of one or more Indians.

[8] Eleven states (Alaska, California, Colorado, Louisiana, Montana, New Mexico, North Dakota, Oklahoma, Texas, Utah, and Wyoming) and seven tribes (Blackfeet Nation, Jicarilla Apache Tribe, Navajo Nation, Shoshone and Arapaho Tribes, Southern Ute Indian Tribe, Ute Mountain Ute Tribe, and the Ute Indian Tribe) conducted compliance work under cooperative agreements with MMS in fiscal year 2007.

[9] Indian leases are not subject to the same 7-year time restriction for making monetary demands.

[10] Although we considered the production inspection results from all 27 BLM field offices in these five states, we excluded production inspection results from two BLM field offices where BLM state inspection and enforcement coordinators could not validate production inspection numbers because they felt that the data in BLM's Automated Fluid Minerals Support System—the database used to track production inspections— were unreliable. We excluded one additional BLM field office because it is implementing a pilot project inspection program using different selection and prioritization criteria; therefore, it is not comparable with the other BLM field offices.

[11] To gain a balance of perspectives of how BLM field offices conduct production inspections, we chose a nonprobability sample of five field office locations—Meeker, Colorado; Vernal, Utah; Farmington, New

Mexico; Pinedale, Wyoming; and Buffalo, Wyoming. In choosing these field offices we sought to ensure we visited BLM field offices that represent a range of BLM state office jurisdictional policies. While this nonprobability sample allowed us to learn about many important aspects of production inspections, it was not designed to be representative of all inspection activities at BLM field offices. As such, the findings cannot be generalized to sites we did not visit.

[12] In the *Cobell v. Norton* litigation—now *Cobell v. Kempthorne*—concerning the government's management of Native American trust funds, a U.S. District Court judge, on December 5, 2001, ordered the Department of the Interior to disconnect from the Internet all information technology systems that house or provide access to individual Indian trust data. The last of the systems was allowed to reconnect to the Internet in May of this year.

[13] The Federal Oil and Gas Royalty Simplification and Fairness Act of 1996, Pub. L. No. 104-185, § 5(a) (1996), provides a 6-year adjustment window.

[14] 30 U.S.C. § 1721a.

[15] 30 U.S.C. § 1724(b).

[16] MMS conducts other types of compliance work, such as limited scope compliance reviews, which examine, among other things, one or more of the variables—sales price, sales volume, and royalty rate—of the royalty equation. MMS did not count these reviews towards its performance goal.

In: Royalties at Risk in the Development...
Editor: Thomas Marinelli

ISBN: 978-1-60692-837-0

Chapter 3

ROYALTIES AT RISK: OVERSIGHT HEARING BEFORE THE COMMITTEE ON NATURAL RESOURCES U.S. HOUSE OF REPRESENTATIVES ONE HUNDRED TENTH CONGRESS FIRST SESSION*

U.S. Government Printing Office

The Committee met, pursuant to call, at 11:00 a.m. in Room 1324, Longworth House Office Building, Hon. Nick J. Rahall, II [Chairman of the Committee] presiding.

Present: Representatives Rahall, Christensen, Grijalva, Bordallo, Costa, Kind, Inslee, Baca, Flake, Pearce, Brown, McMorris Rodgers and Lamborn.

STATEMENT OF THE HONORABLE NICK J. RAHALL, II, A REPRESENTATIVE IN CONGRESS FROM THE STATE OF WEST VIRGINIA

The Chairman. The Committee on Natural Resources will come to order and is convening today to conduct a hearing on Royalties at Risk.

In recent years, many of us have witnessed a sorry pattern across the spectrum of Federal agencies. At every henhouse a fox is stationed. The well-connected few are rewarded at the expense of the common folk, and all the while the people's branch has been looking the other way.

The Minerals Management Service exemplifies this troubling pattern. Throughout the last several years, the MMS seems to have drifted further and further into the grasp of the oil and

---

* This is an edited, reformatted and augmented version of U. S. Government Printing Office publication, Serial Number 110-9, dated March 28, 2007.

gas industry. No doubt this occurred in part because of the agency's dual and conflicting roles.

On the one hand the MMS is charged with developing energy resources on our public lands. On the other hand, it is supposed to collect at a fair value payment owed to people for the development of the resources they own. But something has gone wrong. Development has been thriving. Oil and gas drilling has dramatically increased. It is obvious by the sheer number of rigs that are sprouting up all over the West and drilling platforms in the Gulf of Mexico.

The energy development function at the MMS has been running at full throttle, but when it comes to collecting the payment due to the people, the agency has stalled. At best its performance might be described as slipshod, but some argue it is something more sinister.

Further complicating the collection capability of the MMS is the Royalty-in-Kind program, a program over which I have consistently in my years in Congress raised concerns. The RIK program was designed to allow energy companies to pay the government in product rather than in cash, but it has become an elusive mess.

For the MMS, tracking money was hard enough, but following the flow of oil has proven to be a slippery business indeed. By some accounts the Royalty-in-Kind program has served as a giant loophole, allowing wealthy companies to forego fair payment to the public.

To make matters worse, in 2000 the agency further watered down its audit function through a process known as compliance review. Under this regime, fewer and fewer audits are being conducted even as more and more energy is being produced.

Some seasoned auditors tried to raise red flags, but they were ignored or they were pushed aside. As a result, the people back home who are struggling to pay the mortgage, buy the groceries and pay their annual tax bill have been getting tougher treatment from Uncle Sam than do wealthy multinational oil conglomerates.

The MMS by necessity works closely with representatives of the oil and gas industry, but in far too many cases that closeness has taken on the distasteful appearance of coziness. It is becoming clear that the agency has acted to the benefit of the industry and to the detriment of the public.

The recurring questions are: To what extent did it do so deliberately? How can it be fixed? What can be done to prevent it from happening again?

This committee has long neglected its duty to explore these questions. That ends now. From the moment I took over this chairmanship, I said that we would exercise vigorously and fully our constitutional responsibilities of oversight. That is what we begin today.

I thank the witnesses for taking the time to be with us, and I look forward to hearing their testimony. Before I do, I will recognize the acting Ranking Member, the gentleman from New Mexico, Mr. Pearce.

Statement of Stevan Pearce, A Representative in Congress from the State of New Mexico

Mr. Pearce. Thank you, Mr. Chairman. I am excited to join you at today's hearing, Royalties at Risk. Like yesterday's hearing, I must say that the title gives me pause because of the perception that it creates.

We learned yesterday that of the 262 million acres of BLM land, less than five percent of that land is being used for oil and gas production. The other 95 percent of BLM lands has no oil, no gas production, even though treasured resources are there.

Similarly, only 15 percent of the National Wildlife Refuge System has oil and gas activities. Furthermore, not a drop—zero percent—of our parklands permit oil and gas activities. We learned last Congress that less than three percent of our Outer Continental Shelf is being leased for oil and gas production.

The Outer Continental Shelf and our Federal lands aren't generating any royalties, any of the royalties that it could be, because most of it is off limits. Those are the royalties at risk. If we continue to follow the San Francisco energy policy, those are the royalties at risk we should be discussing.

This hearing is a classic example of penny wise and pound foolish. There are billions of dollars of Federal royalties left on the table because more and more of our Federal lands where much of the energy is remains off limits.

For example, a recent Congressional Research Service study regarding the Arctic National Wildlife Refuge, ANWR, estimated the prospective royalty revenue at $36 billion. In addition, corporate income tax revenue from ANWR was estimated at $75 billion.

I don't mean to suggest that we must not demand that the American taxpayer is fairly compensated for existing oil and gas production, but it is a great shame that the truth of our tremendous energy resource and the associated royalties are being hidden from the American public.

I know that today's testimony will include a discussion of the MMS Royalty-in-Kind program. I have been a supporter of collecting royalties in kind because it is simpler to take a straight percentage of a meter reading rather than to work in a system of dueling auditors over oil's production value. Any time you have money involved, you will attract lawyers and auditors like regulations to bureaucrats.

Every grade school child understands the concept of one for me and one for you. Royalty-in-Kind is much simpler than all of this production value dueling and fighting. Simply put, the more plumbing you have, the more ways there are to clog up the drain.

I worked in the oil and gas industry, and the truth is the matter of trying to value oil is not a simple task. It is a commodity whose price is different every day, depending on where you are. The price for our West Texas crude in eastern New Mexico was always valued at less because it had a lesser quality and a higher cost at the refinery.

When prices were low, oil servicemen like me hurt, and when prices were high you knew that you could make your house payment. That price changes daily and by location. To this end, I look forward to hearing from Assistant Secretary Allred regarding MMS' Royalty-in-Kind program and the progress you have made so that we may put an end to dueling auditors.

Again, I look forward to the testimony and discussions. Welcome to all of you, and thank you again, Mr. Chairman.

The Chairman. Do any other Members have opening statements?

[No response.]

The Chairman. OK. If not, we will proceed with the panel. Panel I is The Honorable C. Stephen Allred, Assistant Secretary, Department of Interior, and Mr. Mark Gaffigan, Acting Director, National Resource and Environment, Government Accountability Office.

Gentlemen, we thank you for being with us today. We have your written testimony. It will be made part of the record as if actually read, and you may proceed as you wish.

Statement of Mark Gaffigan, Acting Director, National Resource and Environment, Government Accountability Office

Mr. Gaffigan. Mr. Chairman, Mr. Pearce, Members of the Committee, good morning. I am pleased to be here to discuss the Minerals Management Service's management of Federal royalty revenues collected from energy resources produced from Federal lands and waters.

Federal lands and waters provide a significant amount of revenue and energy to the American people. Most notably, about $10 billion in annual oil and gas royalty revenue is generated from Federal production that supplies about one-third of all the oil and onequarter of all the natural gas produced in the United States. In addition, Federal lands and waters also provide increasingly important renewable resources such as geothermal energy.

M.M.S. has strived to meet the challenging responsibility of managing these resources, but our oversight work has highlighted some problems that require attention. In my testimony today, I would like to touch upon three areas of our royalties work that point to the need for ongoing attention: Royalty relief, royalties-in-kind and geothermal energy royalties.

First, royalty relief. The waiver or reduction of royalties in order to encourage the development of oil and natural gas has been fraught with problems. Specifically, a series of mistakes and legal challenges in implementing relief under the 1995 Deep Water Royalty Relief Act will likely add billions in unanticipated cost for taxpayers.

As has been widely reported, price thresholds, provisions designed to limit royalty relief in the event of the high prices we are experiencing today, were left off of over 1,000 leases issued in 1998 and 1999. Today, the lack of these thresholds has already cost about $1 billion in foregone royalty revenue.

In addition, current estimates by MMS indicate a range of future foregone royalties of between $6.4 and $9.8 billion. Adding to this problem, a current lawsuit is questioning whether MMS even has the authority to establish price thresholds for any of the leases issued under the 1995 Act, thus bringing into question over 2,000 more leases issued in 1996, 1997 and 2000. If the case is lost, this could add billions more in foregone revenue.

However, not every Federal lease provides royalty free oil and gas to producers. In fact, on most leases royalties are actually paid by producers and collected by the Federal government. Unfortunately, determining proper royalty cash payments has a history of costly, administratively difficult problems resulting from disputes and litigation over the value of oil and gas.

As an alternative to collecting royalties and cash, MMS can choose to accept a portion of the actual oil and gas produced to sell itself, known as taking royalties-in-kind or RIK. We reviewed the RIK program in 2003 and 2004 and found shortfalls in data and information systems that limited the ability to determine how well RIK was doing.

We made recommendations to improve RIK, and MMS was very responsive in implementing them. However, the RIK program has grown significantly, with MMS receiving one-third of its revenue from RIK in Fiscal Year 2005. This raises questions about the ability to effectively manage RIK at these much higher levels. At the request of Congress, we are about to begin an updated look at RIK, ensuring that this important program has continued oversight.

Finally, while MMS has faced significant challenges in managing oil and gas royalties, other energy resources also demand attention. For example, our 2006 review of geothermal royalties found that about 40 percent of MMS royalty data for the geothermal projects we looked at was either missing or erroneous. As with oil and gas, the lack of good data and information limits the ability of MMS to fulfill its royalty management responsibilities.

M.M.S. is charged with an important responsibility in balancing the goals of developing Federal energy resources while ensuring a fair return for the American people. Unfortunately, royalty management problems have created a sea of uncertainty about future development and a crisis of confidence for the American taxpayer that overshadows both goals.

As we continue our work, GAO looks forward to assisting the Congress and MMS to ensure that royalty management continues to have oversight attention.

This concludes my opening remarks. I have submitted a written statement for the record, and I welcome any questions you might have. Thank you.

[The prepared statement of Mr. Gaffigan follows:]

Statement of Mark Gaffigan, Acting Director, Natural Resources & Environment, U.S. Government Accountability Office

Mr. Chairman and Members of the Committee:

We are pleased to be here today to discuss our recent work on the administration of revenues collected from the production of fossil and renewable energy resources on federal lands and within federal waters. Companies that develop these resources do so under leases which generally require the payment of royalties on the resources extracted and produced. These leases are administered by the Minerals Management Service (MMS), an agency within the Department of the Interior (Interior). These resources include geothermal, coal, and, most notably, oil and natural gas (hereafter oil and gas).

In particular, fossil energy resources from federal lands and waters are a critical component of the nation's energy portfolio, supplying more than a third of all the oil and nearly a quarter of all the natural gas produced in the United States in Fiscal Year 2005. Oil and gas companies received over $77 billion from the sale of oil and gas produced from federal lands and waters in Fiscal Year 2006, and these companies paid the federal government about $10 billion in royalties.

In order to promote oil and gas production, the federal government has at times and in specific cases provided ''royalty relief''— the waiver or reduction of royalties that companies would otherwise be obligated to pay. When the government grants royalty relief, it typically specifies the amounts of oil and gas production that will be exempt from royalties and may also specify that royalty relief is applicable only if oil and gas prices remain below certain levels, known as ''price thresholds.'' For example, the Outer Continental Shelf Deep Water Royalty Relief Act of 1995, also known as the Deep Water Royalty Relief Act (DWRRA), mandated royalty relief for oil and gas leases issued in the deep waters of the Gulf of Mexico from 1996 to 2000. These deep water regions are particularly costly to explore and develop. However, as production from these leases has grown, and as oil and gas prices have risen dramatically in recent years, serious questions have been raised about the extent to which royalty relief has been in the interest of taxpayers. These concerns were brought into stark relief when it was learned that MMS issued leases in 1998 and 1999 that failed to include the price thresholds above which royalty relief would no longer be applicable, making large volumes of oil and natural gas exempt from royalties and significantly affecting the amount of royalty revenues collected by the federal government. Further royalty relief is currently available under other legislation and programs, raising the prospect that the federal government may be forgoing additional royalty revenues. Recently, congressional committees, Interior's Inspector General,[1] public interest groups, and the press have questioned whether our nation's oil and gas royalties are being properly managed and whether the oil and gas industry is paying a fair share of revenue to the public resource owners, especially in light of high oil and gas prices, record industry profits, and the daunting current and long-range fiscal challenges facing our nation. GAO has expressed similar concerns, and the U.S. Comptroller General has highlighted royalty relief as an area needing additional oversight by the 110th Congress.[2]

The MMS is authorized by Congress to collect royalties ''in value,'' as a fraction of the revenues companies receive from sale of oil and gas produced on federal leases, or ''in kind,'' as a fraction of the oil and gas that the MMS then sells to recover the government's share of oil and gas revenue. With regard to oil, while MMS has long received relatively small amounts of oil in kind for specific purposes, such as in a past program that provided royalty oil to small refiners at subsidized prices, the bulk of royalties have historically been collected in value. In recent years, however, MMS has taken a growing proportion of oil royalties in kind. Much of this oil was then exchanged for other oil that was put into the nation's Strategic Petroleum Reserve, over 700 million barrels of publicly held crude oil that is stored to ensure emergency supplies in the event of a significant disruption in the normal oil supply. Under the Energy Policy Act of 2005, MMS is charged with ensuring that the revenues it receives when it sells oil taken in kind are at least as great as the revenues it would have received had it taken the royalties in value. The recent expansion of the royalties in kind (RIK) program has raised the obvious question of whether or not this condition is being met.

While fossil energy resources are significant, the federal government also manages royalties from renewable sources such as geothermal energy. Geothermal energy is a unique renewable energy resource in that it can provide a consistent and uninterrupted supply of heat and electricity. Companies drill wells to bring the geothermal fluids and steam to the surface, separate the steam from the fluids as their pressure drops, and use the steam to spin the blades of a turbine that generates electricity. The electricity is then sold to utilities in a manner similar to sales of electricity generated by hydroelectric, coal-fired, and gas-fired power plants, and the companies pay royalties based on the electricity sold.

Due, in part, to increasing demand for electricity, interest is increasing in developing geothermal energy resources as an alternative form of generation. Because many areas that have the potential to produce additional geothermal energy are located on federal lands, the federal government will continue to be a major participant in the future development of geothermal energy. MMS collects the federal geothermal royalties and disburses to the state and local governments its share of these royalties. In 2005, the most recent year for which data are available, MMS collected $12.3 million in geothermal royalties, almost all of which was derived from the production of electricity.

You asked us to provide information from our recent work on the administration of federal royalty revenues at MMS. My testimony today (1) updates our work regarding the fiscal impacts of royalty relief for leases issued under the Deep Water Royalty Relief Act of 1995; (2) describes our recent work regarding the administration of the royalties in kind program, as well as ongoing work on this and related issues we have undertaken for congressional requesters; and (3) provides information on the challenges to collecting and managing geothermal royalties that we identified in recent work.

To address these issues, we relied on recent GAO reports related to MMS's royalty collection systems for oil, gas, and geothermal resources. As part of our ongoing work, we also reviewed the methodology and assumptions used by MMS to produce their February 2007 estimate of foregone oil and gas royalties. Our work follows the issuance of our report last year explaining why oil and gas royalties have not risen at the same pace as rising oil and gas prices.[3] Our work was conducted in accordance with generally accepted government auditing standards.

In summary we found:

- The absence of price thresholds in leases issued in 1998 and 1999 has already cost the government about $1 billion and MMS's most recent estimate indicates a range of future foregone royalties of between $6.4 billion and $9.8 billion over the lifetime of the leases. However, because there is considerable uncertainty about future oil and natural gas prices and production levels, actual foregone royalties could end up being higher or lower than MMS's estimates. MMS is currently negotiating with oil and gas companies to apply price thresholds to future production from the 1998 and 1999 leases. To date, the results of these negotiations have been mixed—only 6 of the 45 companies involved have agreed to terms. Moreover, a pending legal challenge to Interior's authority to include price thresholds on any leases issued under the DWRRA could, if successful, cost the government billions more in refunded and foregone revenue.
- In our most recent audit of the RIK program, conducted in 2004, we found that MMS had not collected the necessary information to determine whether or not the revenues received from its sales of royalty oil were equivalent to receiving royalties in value, largely because it had not developed information systems to rapidly and efficiently collect this information. We made recommendations to the Secretary of the Interior that the agency has implemented and that have improved the administration of the program as it existed at the time. However, the continued expansion of the program raises additional questions about the adequacy of the agency's overall management practices and internal controls to meet the increasing demands of the program. Accordingly, at the request of Congress, we are undertaking a follow-on review assessing, among other things, the agency's ability to quantify and compare administrative costs and revenues of the RIK and royalties in value programs and the extent to which the revenues collected under the RIK program are equal to or greater than what would have been received had they been taken in value.
- In a 2006 report on geothermal royalties, we found that MMS had erroneous and missing historical geothermal royalty data and did not collect sufficient data from royalty payors to accurately asses whether MMS was collecting the amount of royalties required by statute. The Energy Policy Act of 2005 included provisions that significantly changed how geothermal royalties are calculated but also instructed the Secretary of the Interior to seek to maintain the same aggregate level of royalties over the next ten years that would have been collected prior to

the Act's passage. We found that in order to compare royalties collected under the provisions of the Act with what would have been collected under the old system would require historical data on gross revenues from geothermal electricity sales as well as accurate royalty data. However, we found that MMS did not have sufficient historical gross revenue data with which to establish a baseline for past royalties paid as a percentage of electricity revenues. Further, about 40 percent of MMS's royalty data was either missing or erroneous for the projects we reviewed. In our report we recommended that the Secretary of the Interior direct MMS to correct these deficiencies and the agency agreed with our findings and recommendations. We are continuing to monitor the agency's efforts to address these shortcomings.

Background

Interior oversees and manages the nation's publicly owned natural resources, including parks, wildlife habitat, and crude oil and natural gas resources on over 500 million acres onshore and in the waters of the Outer Continental Shelf (OCS). In this capacity, Interior is authorized to lease federal fossil and renewable energy resources and to collect the royalties associated with their production. These substantial revenues are disbursed to 38 States, 41 Indian Tribes, Interior's Office of Trust Funds Management on behalf of some 30,000 individual Indian royalty owners, and to U.S. Treasury accounts.

Royalties paid for fossil and renewable resources extracted from leased lands represent the principal source of the $12.6 billion in revenues managed by MMS' $10.7 billion, more than 85 percent of revenues received in Fiscal Year 2006.[4] Of these, oil and natural gas leases are the most significant component of royalties, composing on average nearly 90 percent of the royalties received over the past five years. For oil and gas, production royalties are paid either in value or in kind. The OCS Lands Act of 1953, as amended, and the Mineral Leasing Act of 1920, as amended, authorize the collection of production royalties either in value or in kind for federal lands leased for development onshore and on the OCS. Furthermore, according to MMS, the terms of virtually all federal oil and gas leases provide for royalties to be paid in value or in kind at the discretion of the lessor. The Energy Policy Act of 2005 provides additional statutory requirements to support the operation and funding of a program for managing federal oil and gas royalties in kind.

Additionally, MMS also collects revenue generated by exploration and development of geothermal energy resources commonly used to generate electricity.[5] Until recently, the Geothermal Steam Act of 1970, as amended, directed MMS to disburse royalties collected from geothermal energy development such that 50 percent of geothermal royalties be retained by the federal government and the other 50 percent be disbursed to the states in which the federal leases are located.[6] A provision of the Energy Policy Act of 2005 changed the distribution of the royalties collected from geothermal resources. While 50 percent of federal geothermal royalties must still be disbursed to the states in which the federal leases are located, an additional 25 percent must be disbursed to the counties in which the leases are located, leaving only 25 percent to the federal government.

Billions of Dollars of Royalty Revenue will be Foregone because of Problems Associated with Royalty Relief

As Assistant Secretary Allred of Interior recently testified before the Congress, the absence of price thresholds in leases issued in 1998 and 1999 has already cost the government almost $1 billion and MMS has estimated a range of potential future foregone revenue for these leases of between $6.4 billion and $9.8 billion. MMS calculated these estimates under a range of assumptions about oil and natural gas prices and future production levels. We reviewed MMS's assumptions and methodology for estimating the potential foregone revenue from 1998 and 1999 leases and found them to be reasonable. However, because there is considerable uncertainty about future oil and natural gas prices and production levels, actual foregone royalties could end up being higher or lower than MMS's estimates.

MMS is currently negotiating with oil and gas companies to apply price thresholds to future production from the 1998 and 1999 leases. If successful, this approach would partially undo the omission of price thresholds for future production, thereby implementing the royalty relief as though price thresholds had been included in the leases. However, the results of the negotiation have been mixed so far—as of late February 2007, only 6 of 45 companies have agreed to terms, and a current legal challenge to Interior's authority to set price thresholds on any DWRRA leases may further deter or complicate a negotiated settlement.

In addition to forgone royalty revenues from leases issued in 1998 and 1999, royalty revenues on leases issued under DWRRA in 1996, 1997, and 2000 are also threatened pending the outcome of a legal challenge regarding price thresholds. Specifically, Kerr-McGee filed suit against the Department of the Interior in early 2006, challenging its authority to place price thresholds on any of the leases issued under the DWRRA. In effect, this suit seeks to remove price thresholds from the leases in question. In June 2006, Kerr-McGee agreed to enter into mediation with Interior in an attempt to resolve the issue; however, the mediation was unsuccessful and litigation has resumed. As of March 2007, the leases in question have generated approximately $1 billion in royalties. If the government loses this legal challenge, it may be required to refund these royalties—perhaps with interest penalties—and to forego any future royalties on these leases, and perhaps any lease issued during 1996, 1997, and 2000. As a result, the government could stand to lose billions of additional dollars.

The RIK Program Has Been Unable to Demonstrate Its Effectiveness Due to Data Limitations

We reviewed the RIK pilot program for this committee in two separate reports in 2003 and 2004 and found that MMS did not collect the necessary information to effectively monitor and evaluate the program.[7] This information includes the administrative costs of the RIK program and the revenue impacts of all sales. We found that MMS lacked this information largely because it had not developed information systems to rapidly and efficiently collect this information.

We made several recommendations in our 2003 and 2004 reports to address the shortcomings we identified. Specifically, to further the development of management controls for MMS's RIK program, we recommended that the Secretary of the Interior instruct the appropriate managers within MMS to identify and acquire key information needed to monitor and evaluate performance prior to expanding the RIK program. We specified that such information should include the revenue impacts of all RIK sales, administrative costs of the RIK program, and expected savings in auditing revenues. We also recommended that MMS clarify the RIK program's strategic objectives to explicitly state that the goals of RIK include obtaining fair market value and collecting at least as much revenue as MMS would have collected in cash royalty payments. MMS agreed with both recommendations and has taken several steps to address these shortcomings.

We acknowledge the agency's efforts and, within the context of the program's scope at the time of our report, consider our recommendations implemented by the agency. However, the expansion of use of RIK since our last review raises an additional concern. The RIK program has actively expanded the scope of its operations as MMS has increasingly opted to take royalties in kind rather than in cash. As MMS reported in its September 2006 Report to Congress, today's RIK operation manages a significant portfolio of the nation's oil and gas royalty assets collected primarily from federal leases in the Gulf of Mexico. This portfolio has expanded more than three-fold from 1999 to present—some 82 million barrels of oil equivalent were exchanged in kind in Fiscal Year 2005—and is expected to continue to grow for the foreseeable future. The Energy Policy Act of 2005 permanently established an RIK operation with administrative and business costs to be paid from royalty revenues generated by RIK sales, effectively transitioning the program from pilot status to a steady-state business operation and potentially enabling a further expansion of the RIK program. The Act restricts the use of RIK to those situations where the benefit is determined to equal or exceed the benefit from royalties in value prior to the sale. However, the larger scale of the RIK program at present makes it unclear that MMS can effectively and accurately make this determination going forward.

Noting this issue, we are undertaking work for the Congress. Specifically, we have several ongoing reviews assessing, among other things, MMS's ability to quantify and compare administrative costs and revenues of the RIK and royalties in value programs; the effectiveness of the systems used to collect, account for, and disburse royalties; and the accuracy of royalty revenue collection, including evaluating whether the value of RIK payments equal or exceed the value of royalties that would have been received in value for oil and gas as required by statute.

MMS Does Not Collect the Data Necessary to Assess Whether Geothermal Royalties Remain Constant as Required by Law

In a 2006 report on geothermal royalties, we found that MMS had erroneous and missing historical geothermal electricity revenue data and did not collect sufficient data from royalty payors to accurately asses whether MMS was collecting the amount of royalties required by statute. [8] Specifically, about 40 percent of the royalty revenue data for royalty payors was either missing or erroneous in the projects we reviewed. In addition, MMS did not have sufficient historical gross revenue data for geothermal electricity sales.

MMS is charged with collecting and distributing royalties collected from the development of geothermal resources used to generate electricity. The Energy Policy Act of 2005 included provisions that significantly changed how geothermal royalties are calculated but also instructed the Secretary of the Interior to seek to maintain the same level of royalties over the next ten years that would have been collected prior to the Act's passage. We found that to meet the statutory requirements, MMS will need to calculate the percentage of gross sales revenues that lessees will pay in future royalties from electricity sales and compare this to what lessees would have paid prior to the Act. In order to compare royalties collected under the provisions of the Act with what would have been collected under the old system would require historical data on gross revenues from geothermal electricity sales as well as accurate royalty data on those sales.

As a result of the insufficient gross revenue data and missing or erroneous royalty revenue data, MMS is unable to determine if it is collecting the amount of royalties on geothermal electricity production as required in statute. In our report we recommended that the Secretary of the Interior direct MMS to correct these deficiencies and the agency agreed with our findings and recommendations. We will continue to monitor the agency's efforts to address these shortcomings.

Conclusions

As seen by all the attention royalties management has received in the Congress and the media, Interior's performance in managing this effort is a cause for concern. Billions of dollars have been lost already and potentially billions more are at risk. In a time of dire long-term national fiscal challenges it is urgent that this problem be fixed and the confidence of the American public that the sale of its national resources is generating a fair return be restored. Our work on this issue is continuing on multiple levels, including comparing the value of royalties taken in kind to the value of royalties taken as cash, reviewing the diligence of resource development, and evaluating the accuracy of the agency's cost, revenue, and production data.

We look forward to this continued work, and to helping this committee and the Congress as a whole exercise oversight of this important issue. Mr. Chairman, this concludes my prepared statement. I would be pleased to respond to any questions that you or other members of the Committee may have at this time.

GAO Contact and Staff Acknowledgments

For further information about this testimony, please contact me, Mark Gaffigan, at 202-512-3841 or gaffiganm@gao.gov. Contact points for our Offices of Congressional Relations and Public Affairs may be found on the last page of this statement. Contributors to this testimony include Frank Rusco, Assistant Director; Robert Baney; Ron Belak; Philip Farah; Doreen Feldman; Glenn Fischer; Dan Haas; Chase Huntley; Dawn Shorey; Barbara Timmerman; Maria Vargas; and Jacqueline Wade.

Related GAO Products

- *Oil and Gas Royalties: Royalty Relief Will Likely Cost the Government Billions, but the Final Costs Have Yet to Be Determined*, GAO-07-369T (Washington, D.C.: Jan. 18, 2007).

- *Suggested Areas for Oversight for the 110th Congress*, GAO-07-235R (Washington, D.C.: Nov. 17, 2006).
- *Department of Interior: Royalty-in-Kind Oil and Gas Preferences*, B-307767 (Washington, D.C.: Nov. 13, 2006).
- *Royalty Revenues: Total Revenues Have Not Increased at the Same Pace as Rising Natural Gas Prices due to Decreasing Production Sold*, GAO-06-786BR (Washington, D.C.: June 21, 2006).
- *Renewable Energy: Increased Geothermal Development Will Depend on Overcoming Many Challenges*, GAO-06-629 (Washington, D.C.: May 24, 2006).
- *Mineral Revenues: Cost and Revenue Information Needed to Compare Different Approaches for Collecting Federal Oil and Gas Royalties*, GAO-04-448 (Washington, D.C.: Apr. 16, 2004).
- *Mineral Revenues: A More Systematic Evaluation of the Royalty-in-Kind Pilots is Needed*, GAO-03-296 (Washington, D.C.: Jan. 9, 2003).

ROYALTIES COLLECTION

Ongoing Problems with Interior's Efforts to Ensure a Fair Return for Taxpayers Require Attention

What GAO Found

The absence of price thresholds in oil and gas leases issued by MMS in 1998 and 1999 has already cost the government about $1 billion and the agency has recently estimated that future foregone royalties would be $6.4 billion to $9.8 billion over the lives of the leases. Precise estimates of the actual foregone royalties, however, are not possible at this time because future projections are sensitive to price and production levels, both of which are subject to change. MMS is currently negotiating with oil and gas companies to apply price thresholds to future production from these leases, with mixed results—only 6 of the 45 companies involved have agreed to terms. Moreover, a pending legal challenge to Interior's authority to include price thresholds on any leases issued under the Deep Water Royalty Relief Act could, if successful, cost the government billions more in refunded and foregone revenue.

In our most recent review of the royalty in kind (RIK) program, conducted in 2004, we found that MMS was unable to determine whether the revenues received from its sales of oil taken in kind were equivalent to receiving royalties in value, largely because it had not developed systems to rapidly and efficiently collect this information. We made recommendations that the agency has implemented that have improved the administration of the program as it existed at the time of our report. However, the continued expansion of the program raises a new question about the adequacy of the agency's overall management practices and internal controls to meet the increasing demands placed on the RIK program. Accordingly, we are undertaking follow-on reviews assessing, among other things, the agency's ability to quantify and compare administrative costs and revenues of the RIK and royalties in value programs and the extent to which the revenues collected under the RIK program are equal to or greater than what would have been received had they been taken in value.

In a 2006 report on geothermal royalties, we found that missing and erroneous historical data, as well as insufficient data on electricity sales, meant that MMS is unable to accurately determine whether it was collecting royalties as directed by statute. The Energy Policy Act of 2005 included provisions that significantly changed how geothermal royalties are calculated but also directed Interior to maintain the same level of royalties over the next ten years that would have been collected prior to the Act's passage. We found that making this determination requires historical data on sales of electricity produced from geothermal resources as well as accurate royalty data. However, MMS did not have sufficient historical gross revenue data with which to establish a baseline for past royalties paid as a percentage of electricity revenues. Further, about 40 percent of MMS's royalty data was either missing or erroneous for the projects we reviewed. We recommended that MMS correct these deficiencies and the agency agreed. We are continuing to monitor the agency's efforts.

Why GAO Did This Study

The Department of the Interior's Minerals Management Service (MMS) is charged with collecting and administering royalties paid by companies developing fossil and renewable energy resources on federal lands and within federal waters. To promote development of oil and natural gas, fossil resources vital to meeting the nation's energy needs, the federal government at times has provided ''royalty relief'' waiving or reducing the royalties that companies must pay. In these cases, relief is typically applicable only if prices remain below certain threshold levels. Oil and gas royalties can be taken at MMS's discretion either ''in value'' as cash or ''in kind'' as a share of the product itself. Additionally, MMS also collects royalties on the development of geothermal energy resources—a renewable source of heat and electricity—on federal lands.

This statement provides (1) an update of our work regarding the fiscal impacts of royalty relief for leases issued under the Deep Water Royalty Relief Act of 1995; (2) a description of our recent work on the administration of the royalties in kind program, as well as ongoing work on related issues; and (3) information on the challenges to collecting geothermal royalties identified in our recent work.

To address these issues we relied on recent GAO reports on oil, gas, and geothermal royalty collection systems. We are also reviewing key MMS estimates and data.

[The response to questions submitted for the record by Mr. Gaffigan follows:]

April 27, 2007
The Honorable Nick J. Rahall
Chairman
Committee on Natural Resources
House of Representative

Dear Representative Rahall:

This letter acknowledges the questions submitted by the Committee concerning our testimony on royalties at risk before the House Committee on Natural Resources on March 28, 2007. Please see the enclosure for our responses.

Sincerely yours,

Mark Gaffigan
Acting Director
Natural Resources and Environment

Enclosure

Question 1: Benefits of Deep Water Royalty Relief Act—What benefits did the country receive by encouraging exploration and development of oil and gas in deep water through enactment of the Deep Water Royalty Relief Act of 1995 during a time of very low commodity prices? In your answer, can you quantify corporate and individual income tax revenue as well as job impacts?

GAO Response—A number of Congressional members asked GAO to examine the costs of royalty relief—in light of rapidly rising oil and gas prices, press reports, and questions by other interested parties as to whether the oil and gas industry was paying its fair share of royalties. These members did not ask us to examine the benefits. However, GAO has indicated in past testimony that benefits are an important part of an overall assessment of royalty relief. These benefits may include increased bonus bids, greater production, increased oil and gas exploration, greater employment in the oil and gas industry, and increased tax revenues. An accurate estimate of these benefits arising from royalty relief would be difficult to achieve because one would need to determine how much of the oil production, exploration, and employment that has occurred is the result of, rather than simply coincident with, royalty relief. Higher oil prices since passage of the act have likely led to increased production, more exploration, and greater employment, even in the absence of royalty relief.

Question 2: Geothermal Recommendations—You indicated that MMS agreed with your recommendation included in your 2006 geothermal study. Are you satisfied that they followed up on your recommendation?

GAO Response—Shortly after release of our report, MMS indicated plans to implement our two recommendations. These recommendations involved requiring payors to report gross sales revenues and requiring MMS to correct erroneous or missing data as necessary. In July 2006, MMS published revised geothermal valuation regulations, in accordance with the Energy Policy Act, which was necessary prior to establishing reporting requirements. MMS reported that it incorporated stakeholder comments into the final regulations and that it will publish these regulations soon. MMS is also revising the geothermal payor handbook to reflect new reporting requirements, including our recommendation to report gross sales revenues. With regards to correcting erroneous and missing data, MMS reports that it completed one full audit and sent an order to the payor to submit royalty underpayments for past years. MMS reports that it is conducting compliance reviews on other leases that we identified as having monetarily insignificant underpayments. GAO will continue to monitor and assess the adequacy of MMS efforts to implement these recommendations until MMS completes its work.

The Chairman. Secretary Allred?

STATEMENT OF THE HONORABLE C. STEPHEN ALLRED, ASSISTANT SECRETARY, DEPARTMENT OF THE INTERIOR

Mr. Allred. Thank you, Mr. Chairman, Mr. Pearce, Members of the Committee. I appreciate the opportunity to meet with you today to discuss these issues that are facing the Department of Interior and the Minerals Management Service.

This is my first opportunity to come before the House before I was confirmed six months ago. I thought it was important for us to give you an insight into my background and perhaps will explain some of the actions that I am taking and planning to take.

By way of background, I grew up on a potato farm and ranch in Idaho. This is my first Federal experience. I have over 20 years of experience in state government. Most recently I assisted then Governor Kempthorne in the creation of Idaho's Department of Environmental Quality and became its first director.

In addition, I have spent 20 years in the private sector, and in fact that is the largest continuous portion of my career, where I was involved in engineering and construction. The organization that I ran as a chief executive of that organization did over $600 million a year in business.

At the outset of my testimony I want to state my belief that government employees have an obligation to protect the public interest of the United States, and they need to be perceived as doing so. If the public believes that we somehow have failed in this obligation, whether in fact or perception, that damage is to the detriment of all of us.

As you know, Secretary Kempthorne places great importance on the Department, its agencies and its employees acting in a highly ethical manner both in fact and perception.

What I want to do in the next couple of minutes is tell you what the Secretary and I are doing to deal with the issues that are the subject of this hearing. After the Senate confirmed me as assistant secretary some six months ago, Secretary Kempthorne asked me to review and manage the issues involving the absence of price thresholds in the deep water leases issued in 1998 and 1999 in the Gulf of Mexico and the other royalty management issues.

Regarding royalty management, the Department has reviewed the Inspector General's report of December 2006 concerning the collection of oil and gas revenues. The Department is actively implementing the recommendations and findings of the IG's review.

In addition, I have traveled to the Minerals Management Service's Denver Operations Center where these activities take place and have reviewed their royalty collection process. Earlier this month in fact I personally observed a royalty-in-kind sale.

We have formed a high level panel to look at these processes and procedures. That panel is chaired by two extremely capable individuals, former Senators Bob Kerrey and Jack Garn. Other members include individuals experienced in Federal revenue collection and those of the state and tribal governments.

As with any large organization with complex operations, there are going to be many opportunities to improve those operations. We look forward to receiving the Committee's recommendations and to further improving our activities.

I find a lot of confusion. In fact, I had a lot when I first got involved in the question of royalties. I would like to take just a short period of time to review how royalties are determined.

The first chart that I would like to put up is the Royalty-in-Value program. What I would like to do is to direct your attention to the left-hand part of that chart. The Minerals Management Service supervises the meters on every source of oil and gas and receives reports with regard to the production of those commodities. They also receive reports from the left-hand side of that chart from those who buy the oil and so we have two sources of independent confirmation as to the amount of oil that is produced.

In between we have two actions which cause most of the argument having to do with the royalty-in-value type of royalty. The first is transportation by the producer to the buyer. The second is the processing of that oil since much of the oil contains both oil and water.

Under the system that Congress has determined that we should use, the producers are allowed to deduct that transportation and that processing from the value of the oil that they produce and upon which they have to pay royalties. It is pretty clear what they produce.

It is generally unclear as to what the costs and whether they are at arm's length relationships for the transportation and the processing of those oils. That is where much of the dispute comes about and the reason that we have to audit and review their compliance.

The second type of royalty that I would like to discuss is the royalty-in-kind. Again, I would direct your attention to the left side of that chart because again we have the same meters which are supervised by the Minerals Management Service and which are reported in actually a couple of different forms to the Minerals Management Service to determine the amount of product, oil or gas, that is produced from that source.

In addition, since we now take our share of that oil or gas and we transport it and process it and sell that on a competitive basis, we know what has been produced. We know what it costs. We do not have the arguments that we have with the producers as to what those deductions should be so the advantage to us in the royalty-in-kind over royalty-in-value is that it removes most of the arguments that we have to resolve with the producers with regard to those.

Incidentally, in 2005 the RIK version of royalties generated $32 million more to the Federal treasury than it would if we had taken the same amount of royalties in the traditional Royalty-in-Value program. Because of the reduced need for the detailed audits, we also believe that we avoided almost $4 million in costs that we would have otherwise incurred if we had managed that under the Royalty-in-Value program.

Now also, as you are probably aware, we will begin in July delivering most of the royalty-in-kind oil to the Strategic Petroleum Reserve to fill its existing capacity. We will maintain a small amount of sales to the Small Refiner program and a small amount in the market so that in 2009 when the repository is filled we will still have the capability to take that and deal with that oil.

I want to emphasize. We have the choice of whether we take it in royalty-in-value or royalty-in-kind, and we make that decision based upon an analysis of what is in the best interest of the government.

I am convinced after my reviews that MMS is collecting the royalties that are set forth by the legislation that Congress has enacted. Again, as with any large and complex operation, there are opportunities to do better and to improve our operation. I and the Minerals Management Service intend to do that.

Just some quick statistics I thought you might be interested in with regard to royalty management. In 2006, we collected $12.8 billion. That involved some 380,000 production sources per month, and we collected royalty from approximately 264,000 royalty-bearing points each month. That is from about 29,000 different leases. We have 530 employees involved in that, of which 118 are compliance staff, 135 are auditors, 120 are contract personnel, and 108 are state and tribal audit partners.

Another interesting statistic is in the 2000-2005 period we completed 1,572 audits resulting in the collection of an additional $98 million, 495 compliance reviews collecting an additional $94 million in revenue.

An interesting statistic I think that is important that our audits returned an average of $2.06 per audit dollar spent. Compliance reviews returned an average of $3.27 per dollar spent in those compliance reviews. MMS enforcement actions in the period 2002-2006 have resulted in the collection of over $52 million in penalties.

Regarding the price thresholds in 1998 and 1999, we have reviewed the Inspector General's report of January 2007 and have I believe gained a fairly complete understanding of the information that is available there.

I have also traveled to the New Orleans Regional Office where the oil and gas operations are regulated. I have asked the Minerals Management Service to form a group, including members of the Inspector General's staff, to conduct a lessons learned review of the information that he has and to apply those to our leases.

The Chairman. Mr. Assistant Secretary, can you wrap up here in the next 30 seconds?

Mr. Allred. I sure can. I want to make the point that since 2001 all of the leases that we have issued have included the price thresholds.

I would be most happy to answer any questions that you have.

[The prepared statement of Mr. Allred follows:]

Statement of C. Stephen Allred, Assistant Secretary, Land and Minerals Management, United States Department of the Interior

Mr. Chairman, thank you for the opportunity to appear here today to discuss with you the Department of the Interior's role in managing energy production on the Outer Continental Shelf and revenue from all Federal and Indian mineral leases. I know this Committee has been instrumental in shaping our domestic energy program, particularly with regard to encouraging environmentally sound development of our domestic oil and gas resources on the Outer Continental Shelf.

The Department and its agencies, including the Minerals Management Service (MMS), serve the public through careful stewardship of our nation's natural resources. The Department also plays an important role in domestic energy development. One third of all energy produced in the United States comes from resources managed by the Interior Department.

As energy demand continues to increase, these resources are all the more important to our national security and to our economy. The Energy Information Administration estimates that, despite increased efficiencies and conservation, over the next 20 years energy consumption is expected to grow more than 25 percent. Even with more renewable energy production expected, oil and natural gas will continue to account for a majority of energy use through 2030. Interior's domestic energy programs, particularly offshore oil and gas production, will remain vital to our national energy portfolio for some time to come, as evidenced in Figure A attached at the end of my statement.

Since assuming the duties of Assistant Secretary of Land and Minerals Management six months ago, I have developed a deeper appreciation for the complexities involved in managing federal energy production. I also am committed to ensuring that we provide an accurate and transparent accounting of the revenue this production generates for the American people.

At the direction of Secretary Kempthorne, two important topics have been my major focus over the past six months—the deep water leases issued without price thresholds for royalty relief in 1998 and 1999, and the management of royalty revenues.

I would like to begin by providing some background on MMS's role in Federal energy production and revenue collection. I then will discuss in greater detail the two primary issues I am focusing on with MMS.

Background

The MMS has two significant missions related to energy: managing access to offshore federal energy resources and managing revenues generated by federal and Indian mineral leases, on and offshore. Both of these functions are important to the nation's economic health and are key to meeting the nation's energy needs.

The Federal Outer Continental Shelf (OCS) covers 1.76 billion acres and is a major source of crude oil and natural gas for the domestic market. In fact, according to the Energy Information Administration, if the Federal OCS were treated as a separate country, it would rank among the top five nations in the world in terms of the amount of crude oil and second in natural gas it supplies for annual U.S. consumption.[9]

Since 1982, MMS has overseen OCS production of 11 billion barrels of oil and more than 116 trillion cubic feet of natural gas.

Since 1982, OCS leasing has increased by 200 percent and oil production has increased by 185 percent. According to MMS's calculations, within the next 5 years, offshore production will likely account for more than 40 percent of oil and 20 percent of U.S. natural gas production, primarily due to deep water discoveries in the Gulf of Mexico.

Attached Figure B shows the Energy Information Administration's 2007 forecast for total domestic oil and gas production and illustrates what the significance of the OCS contribution is to the Nation's energy security.

To support increased production offshore, MMS's Proposed 5-Year OCS Oil and Gas Leasing Program for 2007-2012 proposes a total of 21 lease sales.

We are closer to achieving the goals of this proposed program since January, when the President modified a Presidential withdrawal in order to allow leasing in two areas previously closed—the North Aleutian Basin in Alaska and an area in the central Gulf of Mexico. The President modified the leasing status of these two areas in response to Congressional action and the request of Alaska State leaders. In addition, this Administration has increased the royalty rate from 12.5 percent to 16.7 percent for any new deep water leases offered in the Gulf of Mexico.

In implementing the mandates of the Gulf of Mexico Energy Security Act, MMS will offer deep-water acreage in the "181 South" area and in a portion of the Sale 181 area remaining in the Eastern Gulf of Mexico.

Our analysis indicates that implementing the new program would result in a mean estimate of an additional 10 billion barrels of oil, 45 trillion cubic feet of gas, and $170 billion in net benefits for the nation over a 40-year time span.

In addition to providing and managing access to the OCS, MMS administers and enforces the financial terms for all Federal mineral leases, both onshore and offshore and on Indian lands.

These activities have generated an average of more than $9 billion in revenue per year over the past five years, representing one of the largest sources of non-tax revenue to the Federal Government. (In FY 2006, $12.6 billion was collected, and 60 percent of that was from offshore activities).

Since 1982, the MMS has distributed approximately $164.9 billion to Federal, State, and Indian accounts and special funds, including approximately:

- $101.1 billion to the General Fund of the U.S. Treasury;
- $20.4 billion to 38 states;
- $5.2 billion to the Department's Office of Trust Funds Management on behalf of 41 Indian tribes and 30,000 individual Indian mineral owners; and
- $38.2 billion to the Land and Water Conservation Fund, the National Historic Preservation Fund, and the Reclamation Fund.

MMS carries out these responsibilities under statutory mandates and ongoing oversight by Congress, the Government Accountability Office (GAO) and the Department's Office of Inspector General.

I am happy to point out that for the past five years, as part of its annual CFO audit, MMS consistently has received clean audit opinions from the Office of the Inspector General's contracted independent auditing firm.

1998-1999 OCS Leases without Price Thresholds for Royalty Relief

This January, the Department's Office of Inspector General announced its findings on the 1998 -1999 deep water leases issued without price thresholds. The MMS requested this independent review last year. We appreciate the Inspector General's work and would note that the Department and the MMS have undertaken some procedural and organizational changes regarding lease sale packages and instruments in order to strengthen our leasing procedures.

The Department of the Interior shares Congress's frustration that during the previous Administration price thresholds were not included in the 1998—1999 deep water leases. This Administration has included price thresholds in all deep water leases it has issued with royalty relief. The American people own these resources and are entitled to receive a fair return.

The Deep Water Royalty Relief Act of 1995 required deep water leases issued from 1996 - 2000 to include a royalty incentive to allow companies to produce a set volume of oil and gas before they began paying royalties. Since enactment, the deep waters of the Gulf of Mexico have become one of the Nation's most important sources of oil and natural gas. Price thresholds limit royalty relief when oil and gas prices are high. Price thresholds were included in leases before 1998 and after 1999. They were not included in the 1998—1999 leases.

This matter has been a focus of mine since I assumed this position last fall. In an attempt to address the missing price thresholds, we are continuing to discuss this issue with companies in order to obtain agreements to apply price thresholds to the deep water leases issued in 1998 - 1999. To date our efforts have focused on obtaining the much larger royalty amounts to be realized from future production, estimated to be about $9 billion.

To date we have reached agreements with six companies. This is a significant but we need more companies to sign agreements.

I have adopted three basic principles to guide my actions in seeking to resolve this matter. First, our focus will be to negotiate price thresholds in leases prospectively; second, we will not give economic advantage to one company over another; and finally, we will strive to amend these agreements in a way that will minimize litigation risk.

To achieve these principles, the Administration and the Congress must work together. We cannot do this alone.

We know that the House has already addressed this issue legislatively. We appreciate Congress's efforts to encourage companies to agree to pay additional royalties. However, we must be mindful of potential unintended consequences. H.R. 6 could conceivably result in litigation. If legislation addressed future lease sales, and if a judge were to enjoin future lease issuance for a period of time, the resulting impacts would be significant. Litigation could take years to resolve. The MMS has attempted to project what the potential loss of production, revenue and royalties if lease sales were delayed for a three-year period could look like.

Attached Figure C shows for example, for a 3-year delay, production over 10 years would be reduced 1.6 billion barrels of oil equivalent (boe).

Attached Figure D shows for example, the expected cumulative revenue decline over a 10 year period of $13 billion for a 3-year delay.

We all can agree this would not be in the Nation's best interest. The OCS is a significant supplier of oil and gas. We cannot afford major delays in offshore energy production due to unintended consequences.

We look forward to working with Congress on resolving this issue of national interest.

Management of Royalty Revenue

My second focus is the management of royalty revenue collected from Federal and Indian mineral leases. In FY 2006, about 2,600 companies reported and paid royalties totaling $12.6 billion from approximately 27,800 producing Federal and Indian leases.

MMS's mineral revenue processes and procedures are complex and involve implementing myriad statutory authorities and regulations, as well as a complex set of case law from over 50 years of administrative and judicial decisions on Federal royalty matters.

The process begins when companies calculate their payments for royalties owed the Federal government. Royalties are calculated based upon four components: the volume of oil and gas produced from the lease, which is verified by BLM or MMS officials during regular on-site inspections; the royalty rate, which is specified in the lease document; the value of the oil and gas as determined by regulations; and any deductions for the costs of transporting and/or processing the oil and gas production, which are also determined by regulations. Companies are required to report this information and submit their royalty payments to MMS on a monthly basis.

MMS receives reports and payments from payors and accepts them into the accounting system, similar to filings with the Internal Revenue Service. Fundamental accounting processes identify revenue sources, and funds are distributed to recipients as prescribed by law. Interest is assessed on late and/or under payments.

MMS's audit and compliance program assesses whether royalty payments are correct. The types of questions that arise during compliance activities include whether the company reported and paid its royalty on the right volume, royalty rate, and value and whether the company correctly calculated allowable transportation and processing costs. Findings of underpayments are followed by collection of the payment plus interest. Enforcement proceedings range from alternative dispute resolution to orders to pay and penalty actions.

The current compliance strategy uses a combination of targeted and random audits, compliance reviews, and royalty-in-kind property reconciliations. The strategy calls for completion of the compliance cycle within three years of the royalty due date. In Fiscal Year 2006, this strategy resulted in compliance reviews on $5.8 billion in Federal and Indian mineral lease revenues, 72.5 percent of total mineral revenues paid for calendar year 2003.

In recent years, MMS has completed an increased number of audits, doubling the number of audits in the most recent four-year period over the previous four years. From 1998—2001, MMS, State, and Tribal auditors completed 784 audits compared to the 1,572 audits completed from 2002-2005. This increase is partially the result of the effort in 2005 on the part of MMS to close a significant number of old audits as a result of a recommendation from an external peer review of our audit activities. Collections based on audit work fluctuate from year to year. The apparent reductions in collections resulting from compliance efforts from 2001 through 2004 stand in contrast with very large collections in the 1998-2001 period. This anomaly is due to resolution of numerous lawsuits on undervaluation of crude oil and natural gas during the 1998-2001 period. The result of the resolution of these issues was large payments of additional royalties. Because these issues were resolved, no additional large payments were owed in 2002-2005.

The MMS compliance and enforcement program has generated an annual average of more than $125 million for each of the last 24 years. In other words, MMS has collected a total of more than $3 billion dollars in additional mineral revenues since program inception in 1982.

From FY 2003 through FY 2005, for every dollar spent on compliance reviews, MMS has collected $3.27. For every dollar spent on audits, MMS has collected $2.06.

MMS aggressively pursues interest owed on late payments as required by law. In Fiscal Year 2006, MMS issued over 3,800 late payment interest bills and collected a net amount of $7 million.

MMS has authority to use civil penalties in situations where routine compliance efforts have been unsuccessful. During the last 5 years MMS has collected over $23 million in civil penalties resulting from MRM enforcement actions. So far in FY 2007 MMS has issued over $2 million in civil penalty notices that are now in the administrative process. When combined with other MMS enforcement actions during the same time frame, MMS collected a total of $52.4 million.

Last year, while performing reconciliation of volume imbalances, the MMS promptly identified that the Kerr McGee Oil and Gas Corporation had under-delivered royalty gas volumes to MMS's Royalty-In-Kind (RIK) program—at a time of very high gas prices. MMS pursued the issue and collected $8.1 million—based on these high price periods—to resolve the issue.

In December, MMS announced that a bill for over $32 million had been issued to BP America Production Company for additional royalties and interest due identified through audit work of BP's coalbed methane production that occurred in the state of New Mexico.

These day-to-day efforts are just part of MMS's normal course of business. These efforts are not only effective at ensuring compliance, but also beneficial in bringing the appropriate revenues to the states, Indians, and the American public.

I would like to emphasize, however, that although this work is important, our focus is not on numbers of audits or amounts obtained in collections. The real goal is to increase upfront compliance. We measure success in having higher levels of upfront compliance so that companies are making correct payments the first time. Audits act as a deterrent, but we hope that audits will reveal fewer problems as companies increase voluntary compliance.

MMS has taken steps to improve compliance rates in order to achieve this goal. They include the following:

- Clearer regulations—MMS has made significant progress in developing and implementing clearer regulations, eliminating much uncertainty and ambiguity that previously resulted in major findings.
- RIK—MMS is receiving an increasing percentage of revenues through its RIK program and has eliminated many valuation issues for the RIK volumes. During FY 2005, for example, MMS received about one-third of its revenues through RIK.
- More effective compliance strategies—Compliance reviews have allowed MMS to cover more properties than were possible using audits alone, thereby increasing the deterrent effect. This increased presence encourages companies to be more vigilant about proper reporting and payment.

We appreciate the recent report of the Office of Inspector General concerning the audit and compliance program. The results are similar in substance to audits I have reviewed in State government or in the private sector. My experience is that in any organization with such large and complex operations, I would expect any performance audit to find opportunities for improvement. MMS has embraced the findings, and has an action plan to address them.

We note the Inspector General's major conclusion that compliance reviews are a useful tool in our program, and we look forward to implementing recommendations to further improve our application of compliance reviews. We submit for the Committee's attention our ''Action Plan to Strengthen Minerals Management Service's Compliance Program Operations'' which documents improvement actions taken and planned in this area.

MMS does not work alone in its efforts to ensure the proper collection of royalties; MMS collaborates with the States and tribes on our compliance and audit activities. In addition, every three years, the federal audit function of MMS is peer-reviewed by an outside independent certified public accounting firm. Most recently, in 2005, the MMS audit program was found to meet all applicable government auditing standards. I am also happy to point out that for the past five years, as part of its annual Chief Financial Officer audit, MMS consistently has received clean audit opinions from the Office of the Inspector General's contracted independent auditing firm.

Having said that, it also is true MMS continues to look for ways to improve its programs, practices and performance. We welcome input from this Committee, the full Congress, the Office of the Inspector General, GAO and the public.

In response to the recent interest regarding the accuracy and effectiveness of the MMS's royalty management program, Secretary Kempthorne and I determined that an independent panel should be convened to review the procedures and processes surrounding MMS's management of mineral revenue. We are committed to ensuring our processes are effective and transparent, and we welcome advice and counsel.

The new panel will operate as a Subcommittee under the auspices of the Royalty Policy Committee, an independent advisory board appointed by the Interior Secretary to advise on royalty management issues and other mineral-related policies.

The Subcommittee on Royalty Management has been asked to review prospectively:

- The extent to which existing procedures and processes for reporting and accounting for federal and Indian mineral revenues are sufficient to ensure that the MMS receives the correct amount.
- The audit, compliance and enforcement procedures and processes of the MMS to determine if they are adequate to ensure that mineral companies are complying with existing statutes, lease terms, and regulations as they pertain to payment of royalties.
- The operations of the Royalty-in-Kind program to ensure that adequate policies, procedures and controls are in place to ensure that decisions to take federal oil and as royalties in kind result in net benefits to the American people.

Appointments to the Subcommittee were made on March 21, 2007. We are pleased that former Senators Bob Kerrey and Jake Garn have agreed to serve as co-chairs of this oversight. Secretary Kempthorne served with them in the Senate and knows firsthand of their highest integrity. The other members of the committee bring a wealth of knowledge to this process. They include representatives from state and tribal governments, industry, academia and revenue collection for the government. We are grateful for their service and look forward to their recommendations.

The Subcommittee will conduct its review over a six-month period and then provide its final findings and recommendations to the full Royalty Policy Committee and the Secretary of the Interior. We will be happy to share the recommendations with you when they are available.

State and Tribal Royalty Audits

As part of its compliance assurance activities, the MMS administers delegated and cooperative audit agreements with eleven States and seven Indian Tribes. The States and Tribes are working partners and an integral aspect of the overall onshore compliance efforts. Tribes perform audits on tribal mineral royalties within their reservation and the States perform audits on Federal leases within their boundaries. The MMS conducts compliance reviews and audits to provide compliance coverage over properties not covered by the States and Tribes.

The Federal Oil and Gas Royalty Management Act of 1982 (FOGRMA) tasks the Department of the Interior with ''utilizing the capabilities of the states and Indian tribes in developing and maintaining an efficient and effective Federal royalty management system.'' Title II of this same statute enabled the Secretary of the Interior to enter into cooperative agreements with states and Indian tribes to carry out inspections and audits on Federal and Indian mineral leases within their respective state or reservation. Under Title II, Section 202, we have the Tribal Cooperative Audit Program; and under Section 205, the State Delegated Audit Program).

Since the first agreement was signed with the State of Wyoming in 1981, MMS has held regular meetings with state and tribal representatives to discuss issues of mutual importance. The relationship among MMS and states and tribes has highlighted partnerships as well as contractual obligations.

Funding for States and Tribes participating in the Section 202 and Section 205 programs was around $9.1 million in FY 2006 and remains level for FY 2007. The MMS continues to explore how to best allocate available budget resources for the 202/205 Program. We have analyzed cost, workload, and risk data to apply ''best business case'' criteria to the funding of this program. The mineral revenues at risk and number of producing leases are used to establish funding allocations among States and Tribes. Other factors, such as program effectiveness and anticipated increases and decreases in revenue activity, are also considered.

To manage compliance coverage of the onshore Federal lease universe within available funds, MMS developed a ''business case'' that uses the number of producing leases and total royalty revenues received by states to allocate resources beginning in FY 2006. The attached table reflects the number of leases and revenues received by states and the MMS funding allocation for FY 2004 through 2007. You will notice that the total amount of funds devoted to the audit function of states has not decreased. However, it is apparent from this analysis that some states were significantly over-funded or under-funded in comparison to others. MMS designed the business case to correct such inequities while maintaining overall program funding.

The MMS had several briefings on this methodology with the Congressional delegations representing impacted states, the Department of the Interior's Office of the Inspector General and the Government Accountability Office. During these briefings, the majority of participants seemed satisfied that our methodology was fair and reasonable.

At an August 2006 meeting in Alaska, MMS announced to its state and tribal compliance partners that we will be working on improving the effectiveness of our joint meetings and that MMS will fund one national meeting annually at a central location, as well as regional and topical meetings as needed. The national meeting will address issues common to all states and tribes. Regional and topical meetings will focus on issues specific to a given region of the country. These meetings will provide additional benefits to all parties and enhance communication among MMS and the delegations. States and tribes have also requested training on specific issues which are difficult to address in a national meeting, but will be an integral part of our regional sessions. For example, once the final rule implementing the geothermal provisions of the Energy Policy Act of 2005 is published, MMS will hold a topical meeting with those states that have Federal geothermal production to provide training on the rule and to coordinate our compliance efforts. Discussing this topic at a national meeting is not productive when very few states and no tribes are affected.

The MMS will continue its practice of coordinating with state and tribal delegations in preparing the agenda. For many years, representatives from the Office of the Inspector General have regularly attended the STRAC meetings, and they will continue to be invited.

Conclusion

In the six months since I was confirmed to this position, I have been working closely with the MMS to understand the complex processes associated with accounting for the revenues generated from oil and gas development on Federal lands, including the Outer Continental Shelf. In an effort to gain a greater understanding of this work, I have traveled to MMS's Denver office where I reviewed the procedures and controls used to ensure that minerals revenues are properly reported and accounted for and most recently I attended a sale of Royalty-in Kind oil and gas. I also have visited offices and reviewed operations in the Gulf of Mexico Regional Office.

This work is very important and must be undertaken carefully. Equally important, and very important to Secretary Kempthorne and me, is that we conduct business with the highest standards of ethics possible. Making sure we can live up to that standard has been a high priority of mine. I have stressed, and will continue to stress, our obligation to conduct ourselves in accordance with the highest ethical standards and to be accountable for our actions. Moreover, our conduct must be ethical both in fact as well as in perception.

To summarize my remarks today, I want to reiterate I will continue to focus on several key areas of oversight to the Minerals Management Service.

We will issue our 5-year proposed OCS leasing program on time. This is an important plan that addresses national energy security and facilitates the development of critical energy resources now and in the future.

I will continue to seek prospective royalty agreements with the companies that entered into leases issued in 1998 and 1999 that lack price thresholds in order to capture the majority of the revenues the government would have received.

I am pleased at the results of our efforts thus far, but recognize that there is much more work to be done. I look forward to continuing to work with you, the Members of Congress, to address this important issue.

In addition, I will continue to work with MMS to review and improve our royalty management programs. I have every confidence that MMS will successfully implement appropriate Inspector General's recommendations and that the review by the soon-to-be finalized royalty policy subcommittee will provide a fresh perspective on royalty management issues and challenges.

I welcome your input on all of these initiatives, and I look forward to working with you.

Mr. Chairman, this concludes my remarks. I would be happy to answer any questions you have.

[Note: Attachments to Mr. Allred's statement have been retained in the Committee's official files.]

The Chairman. Thank you both for your testimony this morning.

Let me begin with you, Assistant Secretary Allred. I appreciate the facts and the figures that you have given in your testimony this morning, but, as I referred in my opening statement, it is a fact that since the year 2000 there has been a dramatic decline in the auditing function in favor of compliance reviews.

The IG reported last December that MMS is now using traditional audits on less than 10 percent of leases. The IG also noted that compliance reviews did not provide the same level of assurance as an audit. As I said earlier, in my view there is a dysfunctional situation here ripe for the type of abuses that are repeatedly reported.

So it does appear that we are getting ripped off, plain and simple. You have said here today that you propose additional audits to fix it. I believe that is what you said, and that is my question. Are you going to increase the number of audits that you undertake?

Mr. Allred. Mr. Chairman, one of the advantages we have with the compliance reviews, and incidentally this is not different than what the IRS does as well, is it allows us to look at a much broader spectrum of those who are responsible for providing the royalties.

With the compliance review program, we were able to review about 72 percent. If we were to——

The Chairman. My question is will you increase the number of audits? Yes or no.

Mr. Allred. What we are doing with regard to the audit program is based upon the recommendations of the Inspector General is we are changing our compliance review program to make our audits more risk-based. That may result in more audits, depending upon what comes out of the compliance review.

We are going to have to use both if we are going to have to cover the majority of the royalties that we——

The Chairman. I am not agreeing that both need to be used, but you are saying only that you may increase the number of audits?

Mr. Allred. Mr. Chairman, we haven't finished the risk-based management criteria yet that we are working with the IG on.

The problem here is we have limited resources, as does everyone else. Given the resources we have, we need to try to cover the largest population of royalty payers that we possibly can.

If we do just audits we will cover a much smaller portion of that population, so given both resources what we hope to do, and again I would refer you to the returns per dollar spent. We need to use both. My belief if we need to use both in the appropriate relationships.

The Chairman. Let me ask Mr. Gaffigan. In your testimony you basically state that the MMS is unable to determine whether the royalties it receives in kind are equivalent to receiving those royalties by cash payments. Is that accurate?

Mr. Gaffigan. Yes. When we looked at the program in 2003-2004, and again it was a much smaller program, they were having some difficulty with finding the right information to make that judgment.

We made recommendations to improve the information. We feel that they were responsive to those recommendations. The question is whether today at this larger scale they are implementing those recommendations and making those determinations as we go forward.

That is what we are going to look at in our work for you as we go forward.

The Chairman. So we are basically at a point where it has expanded so rapidly that we need to decide whether it should be eliminated?

Mr. Gaffigan. Well, it is at a good size. I mean, it is one-third of the royalties they took in Fiscal Year 2005. I don't know. Obviously whether you eliminate or not is a policy decision that you guys will make, but it provides an opportunity for them to understand the market.

When we looked at this, we looked at RIK going back to 1999 when it was first talked about. We said at the time there were certain conditions which you needed to look to where RIK may make some sense, and those were sort of the ability to have access to pipelines, processing for gas, large volumes and market expertise.

If you can combine those factors it may make sense to go with an RIK. I think what we will try to do in looking at the RIK program, as well as the RIV program because you can't look at one without the other, is to address those questions.

The Chairman. Thank you.

The gentleman from New Mexico?

Mr. Pearce. I thank the Chairman.

Mr. Allred, on the compliance review process, now the IRS uses compliance reviews and the SEC uses compliance reviews, and yet we are given the impression today that compliance reviews are a complete abomination, that they are something that has only been seen by the robber barons that are currently running the White House.

If you took the full amount of production, and you have that number available to you, don't you? You may not have it today. We know that the amount of percent that the government is supposed to take is about 12.5 percent generally on royalties, but it is creeping up in some places because, frankly, the economics are better.

When you do the calculation, the total number of barrels produced times the royalty rate, you get a number over here at the right-hand side of the page that is some number. Now, I think Johnnie Burton testified that we are getting almost 98 percent or 99 percent. Can you verify that number?

In other words, I am trying to remember the hearing from a couple weeks ago. How much percent of the royalties are we actually collecting if you just do a straight mathematical calculation?

Mr. Allred. Mr. Pearce, I don't have those numbers with me, but as we look through the compliance program what we look to see, and the reason that that number is difficult is, you will remember, you have allowed in the process for the companies to deduct, just like in the income tax——

Mr. Pearce. We are getting way too complex, sir. I appreciate that.

You have made a statement that you are convinced that your agency is collecting what it should set out to collect. If you could get that total number of barrels times the percentage rates on the different blocks because some are different rates, show me the calculation and then show me you said you collect $12.8 billion, show me what it should be because I don't think we are risking much using the compliance review process. Will you get me that?

Mr. Allred. Yes, I will, Mr. Pearce.

Mr. Pearce. OK. All right. Thanks.

In our last hearing about the Clinton leases the IG concluded that this whole business was a mistake by the Clinton Administration, and they based that on several factors, but when we went through that factor the IG himself said that there is no evidence that omission was deliberate, but then on page 7 of his own report says that the Gulf of Mexico supervisor in charge of leases was called by D.C. headquarters and directed to remove the price triggers.

That to me seems deliberate. Does it appear that it was a deliberate act to you?

Mr. Allred. Mr. Pearce, yes, it does.

Mr. Pearce. In the IG report he concluded that the omission was a mistake because there was a policy to include price thresholds. However, are we looked through thousands of emails from that whole period of time we didn't find one instruction to direct people to take. There wasn't one piece of evidence directing that the price thresholds would be included.

He declared then when we made that point, he said well, it was more of an innuendo policy. What do you think about innuendo policies? This is our IG. This is the ranger, the Texas Ranger. What about innuendo policies?

Mr. Allred. Well, Mr. Pearce, I can tell you that in the Department of Interior now we do not have innuendo policies.

Mr. Pearce. All right.

Mr. Allred. We will specify what they are.

Mr. Pearce. The IG also admitted after we began to draw his attention to letters that we held up in front of him and said you say there is no smoking gun, but you happened to leave the smoking guns out. Here are the letters from the Clinton Administration employees that were excluded from his big report, and they all said that the exclusion of price thresholds was intentional.

We are trying to unravel a pretty messy deal here. I don't know. What do you think about the government? Your agency has offered incentives, has even gone beyond, to try to get the companies to kind of voluntarily come into compliance.

I worry about that. I think you all are fair-minded. I don't always agree, but I worry about this one step. Would you talk about that policy where you are encouraging people to change the parameters which they had previously in place?

Mr. Allred. I would be glad to, Mr. Pearce. You know, we believe these are contracts between the Federal government and these companies, and as with any contract you have to have the agreement of both parties to change it.

We are encouraging the companies to come in voluntarily, and six have done so. They represent a little over 20 percent of what we believe will be future production. We are continuing to talk to others to add the price threshold to their leases. We do not believe we can force them, but there are a number of them that believe they need to do so as well.

Those discussions continue. I think though that we will not make further progress in that until Congress decides what role it wants to play because I think currently there is a bit of confusion about what will happen, and the companies are not willing to come in until that is decided.

Mr. Pearce. Thank you.

Mr. Chairman, I have other questions, but I see my time has expired. If we get a second round, I would appreciate it.

The Chairman. OK. The Chair is going to recognize Members by the order in which they came in the hearing this morning. Ms. Bordallo would be next, the gentlelady from Guam.

Ms. Bordallo. Thank you, Mr. Chairman, but I have no questions.

The Chairman. OK. The gentleman from Arizona, Mr. Grijalva. Do you have questions?

Mr. Grijalva. No, sir.

The Chairman. Yes. The Chair will recognize the gentleman from Colorado.

Mr. Lamborn. Thank you, Mr. Chairman. I have no questions, but I would be happy to yield my time to the Ranking Member in case he had any last follow-up questions.

Mr. Pearce. I thank the gentleman.

Mr. Allred, in your testimony you give us some views on H.R. 6. One of the things that legal scholars have talked about is that it appears that H.R. 6 is a takings, a constitutional abridgement; that is, that is offends the Constitution of the United States in taking things without due process or whatever is required.

What do your legal advisors tell you the strength of the Fifth Amendment or the breach of contract claims are on H.R. 6? In other words, what are you expecting if that passes for your challenge to be to implement it?

Mr. Allred. Mr. Pearce, the advice as I understand it is that we believe that that would be challenged in court, and our fear is that that challenge would result in a prohibition or an injunction against going forth with lease sales.

I have some graphs if anyone wants to get more detail, but our concern is that a three-year delay in leasing could well mean about 1.6 billion barrels of oil would not go into the U.S. economy and that about $13 billion over 10 years would not come into the Federal Treasury.

Mr. Pearce. OK. On our next panel we have a witness. Tell me a little bit about the government employees who obtain information in the course and scope of their duties and then their ability to file a claim under the False Claims Act.

What about an MMS auditor that received information from a company in the course of an audit? Should they be able to go under the False Claims Act and file? Can you explain your agency's position on that?

Mr. Allred. I would be glad to, Mr. Pearce. We believe that those auditors should not be able to use information that is available to them through their official course of duties.

There are two different mechanisms if they have concern that they can raise either with management or with the Inspector General. As we have asked the Inspector General to review these issues, and I think you will see a report from him within the next few weeks. I anticipate that that report will indicate that those individuals did not afford themselves either one of those routes to bring their concerns forward.

So I am concerned that auditors within the Department of Interior should not be able to use information derived from their responsibilities as public employees for their own personal gain.

Mr. Pearce. In other words, they have access to data that as a private citizen they could not have access to and so the government requires people to open up their books to show the most sensitive pieces of their company, and then the person who used the government key to unlock that door that would normally be closed then uses data to go out and file a claim in which they personally stand to make a large sum of money.

Now you are just saying that you would disagree with that particular intent. Would you say it crosses an ethical line?

Mr. Allred. Yes, sir, I believe it does.

Mr. Pearce. OK. Talk a little bit about the Tribal Cooperative Audit Program, specific examples compared to what we see in New Mexico or California or North Dakota, if you would. We have just about a minute left, so if you would make it tight.

Mr. Allred. Yes, I will. As you are aware, we contract with states and tribal auditors to assist us in auditing. What we have done recently, and it has caused some consternation, is we have looked at the workload specifically with regard to those states or tribes, and as a result of that we have reallocated some of the funds. We have reduced the funds.

Mr. Pearce. Let me interrupt for just a second. I mean, some of the states are complaining they are not getting enough money, but their amount per lease is tremendously higher than New Mexico at $155 per lease.

You provide a chart of that, and I appreciate that, in your testimony. New Mexico gets $155 per lease. North Dakota, for instance, is up in the $700 range, California above $2,500 per well in this audit process, and yet they are complaining.

We will get into this a little bit later. I see my time has expired. I do appreciate the gentleman.

The Chairman. The gentleman from California, Mr. Costa.

Mr. Costa. Yes. Thank you very much, Mr. Chairman. As a continuation of the hearing we held last month, I thank you for continuing this effort. I think it is very important. This relates to the Committee and the subcommittees' efforts in this 110th Congress.

Mr. Allred, you spoke I guess in your testimony to the Senate Policy Committee on Energy, if I understand it correctly, that there was nothing drastically wrong, and I know you have only been kind of on the job for six months here, but with the Minerals Management Service's bureau, and yet it is my understanding that you also have taken steps toward impaneling a commission to study the Minerals Management Service.

I am trying to understand more clearly the two statements if there is nothing drastically wrong. I mean, we have put together panels when we asked former Secretary Baker and former Congressman Hamilton to make recommendations on the Iraq Study Commission. We have had a similar panel put together when we were looking with Shalala and Dole on issues involving veterans' health care just recently.

If there is nothing drastically wrong, what is the need for the panel?

Mr. Allred. Mr. Costa, I think there is always a need for outside review, and sometimes we get so involved in our day-to-day activities that it is very beneficial to have those outside reviews.

That is why I have supported the Inspector General doing the performance audits. I have had those throughout my career. I think they are very valuable to identify areas where we can continue to improve.

One of my concerns when I asked that panel to be formed by the Secretary was to make sure that we not only deal with the issues and how can we improve, but that we also make sure as we go forth with that that we deal with the perception and the perception that it might not be as well, so in addition to the Inspector General I wanted another group to come in to give us their impression.

Mr. Costa. I appreciate that. So you have appointed the panel, as I understand it. Do you have timelines in terms of when you would like them to report back to the Secretary?

We hope that you will be able to share that with the Committee.

Mr. Allred. Mr. Costa, we would be most happy to. It will be reported to what is called the Royalty Policy Committee, which is a FACA committee. It will be public in that thing, and we would be most happy to brief you.

Mr. Costa. Do you have timelines for them?

Mr. Allred. I have asked them to do it in a six month time period.

Mr. Costa. Good. We will follow up with that.

You made a comment at the end of your first line of questioning to the gentleman from New Mexico, Mr. Pearce, when you were talking about I believe the six out of the 45 that had worked with you, but you said that there was some uncertainty I think—these are my words, not your words—as to the determination as to what would be determined here by the Congress.

Could you explain further? I wasn't clear as to where you were going with that.

Mr. Allred. I would be very happy to. In my continuing discussions with the companies who have 1998 and 1999 leases they are uncertain as to what Congress might do, and because of that they are very reluctant to enter into any further agreements with us until they know what that——

The Chairman. In terms of what, changing the pattern in terms of the royalty? What uncertainty specifically? Can you put your finger on that?

Mr. Allred. I think the questions they have as to whether or not there would be additional costs that might be incurred as a result of what Congress would do.

Mr. Costa. All right. Let me, Mr. Gaffigan, because my time is almost expired.

On the issue of compliance review versus audits, in terms of standard procedure for investigation and accounting practices is there a random sampling or a level in which you think you can best get to the determination as to what in fact is being done appropriately and what is not being done, and do you think in this case we have reached whatever that appropriate randomness is with regard to Minerals Management Service?

Mr. Gaffigan. I would commend to you the IG's report recently in December that looked at this issue. We have been focusing lately on the fiscal impact, but in general when you are doing an audit one of the things, the criteria, you try to look at is materiality; in other words, where are the big dollars, no matter whether it is an audit of royalties or any other issue.

You have to constantly be looking at that. You cannot sort of go with one program and then change as things change. As they move from RIK to RIV or using more RIK, there are also issues to think about. What kind of audit steps do we need to change? What kind of things do we need to look at in RIK?

That doesn't mean, you know, that the audit function goes away just because you have RIK. It is a range of issues. There is no one number that is out there. They should constantly be looking at it, and one of the criteria they should be applying is sort of the materiality of what they are looking at.

Mr. Costa. OK. My time has expired, but I would like to pursue that line of questioning. I may have to submit those questions to you.

I yield back to the Chairman.

The Chairman. The Chair will note the absence of any Member on the Minority side at this time, but will ask unanimous consent that all Members be allowed to submit questions for the record, and we would ask the witnesses to follow up on those written questions at a later time.

The gentleman from Wisconsin, Mr. Kind, is recognized.

Mr. Kind. Thank you, Mr. Chairman, and thank you for holding this very important hearing.

As a former Ranking Member on the subcommittee, a lot of us on the Committee have been disturbed in recent years in regards to some of the reports coming out of MMS and Department of Interior in regards to royalty collections.

I think this hearing is very pertinent and very relevant to the type of work that we need to do in this committee in the coming year, and I appreciate the witnesses' testimony. We are going to have a couple of panels coming before us.

Mr. Allred, let us start with you. Obviously you have heard reports in the media. We have heard reports too in regards to possible retaliation of some auditors in relation to royalty relief and how it is being conducted

If you take a look at those auditors in question, and we are going to have a couple of them testifying in a little bit. If you look at their resume and their background and experience, and they do seem to be eminently qualified, and yet when they were trying to highlight issues and problems at MMS it certainly smacks of retaliation, which is very disturbing.

There may be a very perfect, logical explanation of what took place. Maybe if you can illuminate us a little bit in regards to the cases specifically that would be helpful to the Committee.

Mr. Allred. I would be glad to. First of all, a couple of issues. The first is that there was a process by which those concerns could have been brought forward. They were not.

As I indicated earlier, we have asked the Inspector General to investigate, and I am informed that his report is just about done. That will be available to both you and to us.

Second, the question of retaliation. He has also I believe looked at that, so that information will be available as well. What some are calling retaliation is the issue of whether or not we should continue those auditors to let them audit those accounts for which they would personally benefit from.

The Government Accounting rules prohibit that. They require an independence and so those auditors were removed from that audit function. It was not retaliation. It was done to make sure that we comply with ethics, ethical questions and with the Government Accounting Standards.

Mr. Kind. We will have an opportunity to question them in a little bit in regards to what took place in the individual instances and that, but, getting back to what Mr. Costa was raising, you have testified here today that your Department has been successful in renegotiating six of the deep water leases out of a potential 45 contracts that exist.

Are there any legal challenges right now prohibiting further negotiations with these companies? What is the stalemate with the remaining companies at this point? You indicated there is uncertainty in regards to future congressional action, but is there any other obstacles that MMS is facing right now?

Mr. Allred. There really are none, and we continue those sorts of discussions, but there is a reluctance on their part to agree until they know what the whole playing field is.

As Mr. Gaffigan testified, there is a challenge to the basis of the law that you passed as to whether or not we can apply thresholds, and certainly that is a real question. We believe we are on sound ground in defending your law, but that is a challenge.

Mr. Kind. So that may be freezing the good faith efforts to try to reach an agreement outside of legislative action I assume.

Mr. Gaffigan, let me turn to you. We are just trying to understand the culture at MMS. I have had a chance to briefly review the report that you submitted before the Committee.

In your opinion, what is really the heart of the problem here? Is it just insufficient personnel or resources in data collection there, or is there a greater culture at MMS making it more difficult to collect proper royalty in these circumstances?

It certainly seems in your report that you are concerned about data collection processes and the lack of information in order to make good determinations, but in your opinion that is the real obstacle?

Mr. Gaffigan. I think the challenge comes down to two things: People and the information. MMS is an environment where the kind of people they would hire to do the sorts of things can also get jobs in industry, and when prices are high in the industry and things are going well it is harder and harder to draw the kind of expertise that you need to bring in to do things like royalties-inkind.

The other piece of the puzzle is information. Consistently as we have looked at it as auditors, we have consistently been sort of frustrated by the lack of the ability to get good, timely information.

You know, just in preparing for this hearing we were trying to get a sense of well, how much has gone into RIK lately? You know, right now the best current information is based on Fiscal Year 2005. Here we are in the middle of Fiscal Year 2007. We have seen this theme over and over as we try to look for good information.

Mr. Kind. I see. Thank you. Thank you both for your testimony.

Thank you, Mr. Chairman.

The Chairman. The gentleman from Massachusetts, Mr. Markey.

Mr. Markey. I thank the Chair very much. I am trying to be the winner of the Nick Rahall sound alike contest today. I think I am doing a pretty good job.

The Chairman. You have a long way to go.

Mr. Markey. This pollen is unbelievable today.

Has the Administration, Mr. Allred, reversed its position and now agrees that the language in H.R. 6 would not constitute a takings and would not violate the Equal Protection Clause?

Mr. Allred. Sir, I am not an attorney and I am not aware of any change in our position, nor do I have enough expertise to really comment on whether it would or would not.

Mr. Markey. Well, the testimony from the Administration is that it would constitute a takings and would violate the Equal Protection Clause, so you are saying that hasn't changed.

In your testimony you cite the prospect of litigation as the result of the Administration's opposition to the provisions of H.R. 6. Our committee, Mr. Rahall, I and Mr. Miller and others, we got some of the nation's most respected constitutional scholars to actually help us write the language which is in H.R. 6.

As you know, anyone can bring a lawsuit, but constitutional scholars indicate that this language is in fact going to be upheld so it doesn't really stop obviously an oil company from suing. We assume some will, although it will be on weak constitutional arguments.

Why do you continue to make that case, Mr. Allred, given constitutional experts' views that the language in H.R. 6 is strong and would be upheld if challenged?

Mr. Allred. Congressman, my concern is perhaps a more strategic one from a standpoint of what might happen to the leasing program and what might happen with regard to the amount of oil or the revenues to the United States if there were a challenge and if a company were to be able to get an injunction against the leasing program.

Our analysis has indicated that if we had a three year hiatus as a result of an injunction that it would cost the United States about 1.6 billion barrels of oil and about $13 billion over 10 years to the U.S. Treasury. That is my practical concern.

Mr. Markey. I understand your view on it, but I think that you should understand that that could be the case for any law that this committee passes.

Any law that the National Resources Committee passes that is then signed into law could be brought to court. It is not a reason I think for you to say we shouldn't be passing any laws because oil companies, gas companies, coal companies might take that law to court.

I don't think you should put yourself in the place of the oil industry filing a frivolous lawsuit against an obviously constitutional law which we have already passed through the

House. I think you are siding with the wrong party in this case. I think you should be siding with this Congress and with the American people seeking to reclaim lost revenues rather than with the oil and gas industry.

Now, you stated in your testimony that the Administration will focus on negotiation price thresholds on the leases prospectively. When the companies holding these leases have already received the windfall profit estimated at between $1 and $2 billion from past production, don't you think it makes sense, Mr. Allred, to simultaneously attempt to recover this massive lost revenue as well?

Mr. Allred. Congressman, I believe we should have price thresholds that recover both past and future royalties.

Mr. Markey. What is the Administration doing to recover the already lost royalty payments?

Mr. Allred. One of the things, as I indicated, is that I have tried to look at where I can get the biggest recovery first, and that has been on the future. I have been a little bit handicapped in my discussions.

Mr. Markey. Why is it a mutually exclusive strategy to go after future and not past revenues simultaneously? Why? Are you hamstrung at your agency that you don't have enough personnel? Should we pass an emergency appropriations to get you some more lawyers?

It would be a very small investment for us to have to make to get you another half a dozen lawyers over there to claim $2 billion. I think we would be willing to do it for you. Would you make that request to us?

Mr. Allred. Congressman, my biggest problem has been when I have talked to them and they have said why should we do it because H.R. 6 does not require it prior to October 1.

Mr. Markey. So at that point would you do so?

Mr. Allred. I am sorry. At what point?

Mr. Markey. I know it doesn't require it. Why don't you do it prior to that deadline? I mean, do you need a statutory mandate to do it?

Mr. Allred. I am sorry, but I don't understand. You are saying should we do it irregardless of H.R. 6?

Mr. Markey. Let me just move on to the next question, Mr. Allred.

As part of the renegotiated terms that Interior reached with six of the 45 companies holding these faulty leases, the agreement states that it will terminate if Kerr-McGee prevails in its lawsuit.

What kind of deal is that to agree on a renegotiated contract that still lets the oil companies have the full freedom to sue for more taxpayer money later on?

Mr. Allred. Congressman, I think that is a reality of the legal system. They are challenging your law and whether or not it included the ability for the Department of Interior to place price thresholds.

If the courts find that your law did not include that then we could not prevail in any case to force price thresholds.

Mr. Markey. Did you attempt to get those six companies to waive their rights to file a Kerr-McGee style lawsuit as part of the renegotiation?

Mr. Allred. Congressman, I believe none of those six companies are part of that Kerr-McGee lawsuit.

Mr. Markey. No. As part of your negotiation with them, did you ask them to waive their rights to file a Kerr-McGee style lawsuit?

Mr. Allred. Congressman, no, I did not.

Mr. Markey. You did not. Looking back now, do you think perhaps that should have been something that you had looked at?

Mr. Allred. Congressman, I don't believe they have ever threatened to or have been part of that. I have discussed the Kerr-McGee issue with Anadarko, and I have sought to try to resolve that. I have been unable to do so.

Mr. Markey. Mr. Chairman, thank you. I yield back the balance of my time.

The Chairman. The gentleman from Arizona, Mr. Flake?

Mr. Flake. Thank you, Mr. Chairman.

Mr. Allred, Senator Bingaman in the Senate has said that he doesn't plan to move H.R. 6 because of constitutional questions.

Do you want to comment on that? Does he share the same concerns that you have or that have been expressed by the oil companies?

Mr. Allred. Congressman, I have not had a specific discussion with Senator Bingaman about H.R. 6 so I really can't comment on what his beliefs are.

My concerns have to do with the practical impact of a potential litigation that might deprive us of about 1.6 billion barrels of oil and about $13 billion of income to the Federal Treasury.

What I have sought to do is I have negotiated with these companies to minimize their opportunity to challenge in court what we are doing. I believe they should be paying royalties and the Department of Interior believes they should be paying royalties, but as I have gone forth, and I have learned this through vast experience and a lot of gray hair, is I don't want to do it in a way that might give anybody an opportunity to challenge in court.

What my advice to the Senate and my advice to the House is to however we resolve this issue let us try to do it, and we are most willing to work with all of you. Let us try to do it in a way that does not open the U.S. Government to a challenge of what we do.

Mr. Flake. Do you have any specific recommendations on how to proceed in that regard?

Mr. Allred. Congressman, the Senate asked me that question, and my response was that I believe there is a way, without direct use of subsidies, to encourage more companies to sign by providing the tool to the Department of Interior that would allow us to extend the deep water leases for three years in return for their agreement to include price thresholds. I believe that would bring a number of other companies.

Mr. Flake. Thank you. No further questions.

The chairman. The gentleman from California, Mr. Costa?

Mr. Costa. Thank you.

I appreciate your patience, Mr Allred. Obviously this is something that concerns the Committee. You have appointed—not you but within the Department there was an appointment as it relates to the Interior's Royalty Policy Committee.

As we all know, whether we like it or not in government perceptions are often times challenging. Specifically referencing Mr. Deal, who has been appointed as I guess the vice chair of the Subcommittee on Royalty Management, his previous employment of course with the American Petroleum Institute is one which some would argue might create potential conflicts.

It is important I think for all of us that the Royalty Management Subcommittee and the panel be viewed as independent and that the perception be seen that way. Would you care to comment?

Mr. Allred. Congressman Costa, I would like the opportunity to do so.

The Royalty Review Subcommittee is a subcommittee of the Royalty Policy Committee, which is a FACA committee, and as such it has to have a Royalty Policy Committee member on it. Mr. Deal was the Royalty Policy Committee's choice to be on our subcommittee that we have asked for.

I have little concern, given the other people that are on that, that anybody will bias their report. If you look at Senator Garn or Senator Kerrey or you look at the other people who have been appointed, they do not have backgrounds primarily with business and so I think that what we will get is a very open and a very independent report, and that is what we sought.

Mr. Costa. All right. Thank you.

Mr. Gaffigan, back to the issue of the audits versus the compliance reviews. You spoke about a number of factors that had to be included when Minerals Management Services made those choices.

It is my understanding that the Government Accountability Office is to complete and release an updated report on the Royaltyin-Kind program. Is that correct?

Mr. Gaffigan. We are just beginning that work.

Mr. Costa. OK. So would it be premature then for you to comment or elaborate on some of the findings that are taking place?

Mr. Gaffigan. Yes, because we don't have any findings, but I would say the things we will look at are some of the things we have looked at from the beginning of this program. It goes back to 1999 when we talked about what sort of things need to be out there? What conditions need to be out there to look at RIK?

We identified four remaining factors. We talked about access to pipelines, we talked about access to the processing of natural gas, we talked about having large volumes so the government is in a position of a strong sales position and also market expertise. Those are the sorts of things we will look at.

Following up on that, in our 2003 and 2004 work we asked two basic questions. How do you sort of measure how you would have done against royalties-in-value, and how also have you done in sort of your administrative costs? Do you track litigation costs, and have you any savings there?

When we looked at that in the pilot stage there were some problems in terms of the kind of information that was available to make those judgments. We made recommendations, and we are going to pursue and see how they are doing on a much larger scale.

Mr. Costa. Well, we will want you to keep us updated.

Mr. Gaffigan. Absolutely.

Mr. Costa. I will yield the balance of my time back, Mr. Chairman.

The Chairman. Thank you.

Gentlemen, thank you for being with us today. We appreciate it. Mr. ALLRED. Thank you very much.

The Chairman. Thank you, Mr. Assistant Secretary.

Our next panel is composed of Mr. Bobby Maxwell, Former Auditor, Minerals Management Service, Mr. Kevin L. Gambrell, Indian Land Working Group, and Ms. Ryan Alexander, President, Taxpayers for Common Sense.

The Committee welcomes our panel members, and we will proceed as we normally do. We do have your written testimonies, and they will be made part of the record as if actually read. You may proceed in your individual testimonies as you desire.

Mr. Maxwell?

Statement of Bobby Maxwell, Former Auditor, Minerals Management Service

Mr. Maxwell. Mr. Chairman and subcommittee Members, thank you for the privilege and opportunity to be here today.

I have 26 years' experience auditing oil and gas companies.

The Chairman. Mr. Maxwell, let me ask you to bring the microphone a little closer to you, please.

Mr. Maxwell. OK. Is that better?

The Chairman. Yes, sir. Thank you.

Mr. Maxwell. OK. Thank you.

Mr. Chairman and subcommittee Members, thank you for the opportunity and privilege to be here today.

I have 26 years' experience auditing oil and gas companies. Twenty-two of those years were at Minerals Management Service. I am an expert in auditing sales contracts, revenue and production systems.

I have also received many of the highest awards at MMS for the work I have done. One of those awards included the Meritorious Service Award. I understand auditing and how to protect the assets of the American taxpayer.

M.M.S. has changed over the years. Less auditing is being performed. Less underpaid royalties are being collected. The qualifications of audit staff have decreased. Certain royalty areas are not being audited at all.

Valid orders for royalty payments are sometimes not issued. I personally was not allowed to issue a valid order to Kerr-McGee Corporation for $10 million. I was not allowed to audit royalty-inkind contracts. I was not allowed to order companies to pay interest on late payments.

On the Kerr-McGee issue, I personally filed a false claims lawsuit on behalf of the Federal government. The Department of Justice did not intervene in the lawsuit because it has become a political issue.

On January 23, 2007, a courageous jury in the district court of Colorado found Kerr-McGee Worldwide Corporation guilty of underpaying the Federal government $7.6 million in royalties and also that Kerr-McGee had withheld vital information from the Federal government. With an impartial judge and jury, deliberations only lasted for several hours. The American public has spoken, but now Kerr-McGee is trying for a technical win to not pay the American taxpayer.

Unfortunately, even today MMS states that Kerr-McGee owes no additional royalties. MMS never attended the trial. MMS never reviewed the 60,000 trial documents I received under discovery. MMS never received the thousands of pages of depositions or reviewed them. MMS never reviewed the trial transcripts. MMS really just doesn't know. MMS is the proverbial ostrich that has its head in the sand, sees nothing, knows nothing, but states no royalties are due.

Kerr-McGee personnel's testimony at trial clearly showed they had knowledge that the additional royalty was due MMS. However, their final defense was that MMS never demanded the payment.

M.M.S. must be required to maintain a highly professional and aggressive audit program to collect all the royalties that are due the government. MMS has many qualified auditors, very highly qualified, many with MBAs, CPAs and industry experience. However, many of them are more in clerical or technical positions where they are not using that experience right now.

I think you have reviewed my written testimony, so I thank you for the opportunity to be here and your time and attention, and I will answer any questions you may have.

[The prepared statement of Mr. Maxwell follows:]

Statement of Bobby L. Maxwell, Former Auditor, Minerals Management Service

Mr. Chairman and subcommittee members, thank you for the privilege and opportunity to be here today. This is an opportunity for me to participate in a hearing that will hopefully make the Mineral Management Service (MMS) a better servant of the U.S. taxpayer. MMS is responsible for collecting royalty payments for the U.S. Government and brings in revenue only second to the Internal Revenue Service.

I served the American taxpayer with over thirty years of service, including three years in the U.S. Army. I believe one of the greatest joys in life was the opportunity to work for the government as a citizen of this great country. My only regret is that my career was cut short due to exposing the Federal government's current cozy relationship with the oil and gas industry and its unwillingness to consistently enforce laws and regulations requiring the industry to pay royalties due on Federal oil and gas leases.

I audited the oil & gas industry for over 25 years. I became an expert in reviewing industry contracts, revenue accounting and production systems. I understand professional auditing, industry contracts and how to determine what monies are owed the Federal government under oil and gas leases.

I have a Bachelor of Business Administration degree in accounting from Chaminade University of Honolulu, a Master's in Business Administration from Texas A & M University, and I am a certified public accountant in the states of Oklahoma and Hawaii. I received many awards from MMS for being highly effective as a manager and for audit results. In June 2003, I received the Department of the Interior Meritorious Service Award from the Secretary of Interior, Gail A. Norton.

Around the year 2000, MMS began to change. The auditing function began to be de-emphasized and the enforcement of the lease terms and regulations seemed to become less important. To replace professional audits, top management advocated a new system. This system was called a ''compliance review'' and often resulted in professional auditors being replaced with other staff. The new staff often did not have an educational background containing college level accounting or auditing courses. All senior managers were ''directed'' that they would support the new process as part of their jobs. It was clearly stated that no dissent would be tolerated. I do believe that there is an appropriate need and use for compliance reviews, but they should never be used as a replacement for professional audits.

With the new compliance system we were told not to bother the oil companies. We were told not to be requesting documents as we formerly had with audits. Audit staff was reduced. Many auditors stopped traveling to companies for audits, stopped interviewing oil company staff, stopped visiting marketing departments and field personnel. Audits were marginalized, and accounting and auditing degrees were no longer required. For four years, MMS received a qualified audit report reflecting substandard audit work. In 1992 audits covered 90% of all royalty payments, currently audits and compliance reviews are only covering 72% of royalty payments. Remember that the compliance review is not an audit and does not provide the same level of assurance that royalties were correctly paid. Further, royalty underpayment collections have decreased by over $100 million per year.

By the year 2002, we were no longer allowed to audit certain areas. For example, we were not allowed to review or audit many Royalty-in-Kind contracts. In an attempt to review the contracts and transportation agreements, I was ordered not to get any information from the Royalty-in-Kind division of MMS, or review their contracts for sales of the oil and gas. In an attempt to overcome this major scope limitation on our audits, a meeting was schedule with the Office of the Inspector General and Royalty-in-Kind personnel in Lakewood, Colorado. Audit staff traveled from Oklahoma City, Oklahoma, for the meeting. Neither the Office of Inspector General nor the Royalty-in-Kind personnel showed up for the meeting. I was told not to pursue the issue any further.

This was reminiscent of being directed to not pursue an issue many years earlier. In the early 1990's, I was directed not pursue the issue of oil companies exchanging crude oil by contracts using artificially low exchange values. This resulted in MMS receiving royalties based on a value far below the fair market value of the crude oil. A former employee of ARCO Oil & Gas Corporation filed a False Claims Act lawsuit against the oil companies based on this issue and collected over $400 million for MMS. This was over $400 million that MMS would never have collected on its own initiative.

In 2003, I was chastised for attempting to bill a corporation for interest due on millions of dollars it paid as a result of an audit. However, the audit staff convinced the company to voluntarily remit the interest payment without a bill from MMS. I was told that a system was in place for billing interest and the audit staff should never issue bills for interest relating to royalty underpayments we collected. However, everyone in MMS knew that the system wasn't working and the government was behind many years in the billings and apparently millions of dollars in interest would never be collected. The fact that many millions of dollars in interest would be years late being collected—if ever collected at all—was of no concern to senior management. The Jicarilla Apache Nation complained directly to the Director of MMS and I was allowed to bill all companies for interest due the Jicarilla Apache Tribe.

Every year we were pressured to do less auditing and state that royalties were accurately reported and paid by the oil companies using the compliance review process. In some cases the compliance reviews were adequate and useful in determining if royalties were correctly reported and paid. In other situations, it was only used as a method of smoke and mirrors to state that royalties were in compliance. I was told that finding and collecting royalty underpayments wasn't important, but meeting our Government Performance Results Act standards was what mattered, our operating budget depended upon it. Further, we were directed that MMS would not issue any subpoenas to oil and gas companies for records.

Pressure continued to mount in 2002 & 2003 to not pursue royalty underpayments to the U.S. government by the oil and gas industry. The most well known case is the Kerr-McGee Corporation (Kerr-McGee) royalty underpayments. I developed an order requiring Kerr-McGee to pay MMS an additional $10 million. I was pressured not to issue the order even though it was fully supported by the lease terms and regulations. The pressure came down from the Director of MMS not to pursue these underpayments against Kerr-McGee. I was never told that the order was not supported by the MMS regulations or justified—just not to issue the order. No criteria was ever provided to me stating why the additional royalty should not be collected from Kerr-McGee.

As you know, I filed a False Claims Act lawsuit against Kerr-McGee on behalf on the U.S. government to collect the royalty underpayments by Kerr-McGee. Within days of the lawsuit becoming public, I was notified that I was being terminated from employment with MMS. MMS was determined not to require Kerr-McGee to pay the royalty underpayments. The Department of Justice did not intervene in the lawsuit, but allowed me to take the lawsuit forward on behalf of the American public. This has become a political issue within the Department of Justice.

I personally continued the lawsuit against Kerr-McGee Worldwide Corporation on my own time and at my own expense. It has been a long and difficult road, with absolutely no assistance from the Federal government. On January 23, 2007, a jury in the United States District Court for the District of Colorado found Kerr-McGee guilty of underpaying the Federal government $7,555,886.28, and that Kerr-McGee had failed to disclose to the United States Government all relevant information to determine the value of royalties due. Kerr-McGee's guilt of underpaying royalties and withholding vital information from MMS was quickly determined when the evidence was presented before an impartial judge and jury.

It is important to note that these twelve courageous citizens had no hesitation in finding Kerr-McGee guilty. The overall deliberations were less than four hours, and all indications are that the jurors determined Kerr-McGee was liable in less than two hours into their deliberations. Kerr-McGee's trial evidence did not seriously question the royalty underpayments I had calculated, did not dispute the fact that it had not engaged in reasonable and prudent marketing of this Federal oil, and did not dispute that the buyer, Texon Corporation, L.P. (Texon), had provided other significant consideration, including the assumption of essentially, all of Kerr-McGee's transportation responsibilities. No, Kerr-McGee's primary defense was that MMS had decided not to issue an order to pay.

Kerr-McGee clearly lost in district court, but now is trying for a technical win as a way of not paying its royalties due the American taxpayer.

Not withstanding the fact that the American public has spoken on this issue, even today MMS still states that Kerr-McGee owes no additional royalty. However, MMS never attended the trial; MMS never reviewed the almost 60,000 documents I received under the trial discovery process; MMS never read the thousands of pages of depositions; MMS never listened to the trial testimony or reviewed the trial transcripts. In essence, MMS is the proverbial ostrich with its head in the sand. It sees nothing and hears nothing, but is sure no additional royalty is due. With this type of behavior, it is scary to know that MMS is responsible for protecting the American public's assets and collecting royalties due.

Kerr-McGee's testimony at trial clearly stated that its personnel knew additional royalty was due, but elected not to pay the U.S. Government. Kerr-McGee's accounting and marketing personnel stated that they knew Kerr-McGee received additional value for the oil in the form of services provided by Texon. Further, they testified that they knew royalty was due on this additional value. However, no attempt was ever made to pay MMS the full royalty value. The manager of revenue accounting at Kerr-McGee, Mr. Terry Kyle, even directed an employee to not provide answers to MMS for questions about additional incentives or consideration that Kerr-McGee received in exchange for its sale of the Federal oil to Texon.

Kerr McGee's attorney, Mr. Gorenson, provided an affidavit with respect to one of Kerr-McGee's district court motions indicating that he had full knowledge of the Texon contract terms many years ago. As a Kerr-McGee attorney he made no attempt to have Kerr-McGee pay the proper value of royalties. Rather, he worked to increase Kerr-McGee's profitability at the expense of the American taxpayer.

I personally believe that the knowledge of the contracts and royalty underpayments by Mr. Gorenson and Mr. Kyle are sufficient to show that they knowingly and willfully filed false Federal royalty reports and underpaid the royalties due the American taxpayers. Further, the Director of MMS by stopping a valid order for royalty underpayments makes one wonder if collusion between MMS and Kerr-McGee took place. It is a matter that I personally believe should be closely evaluated. In this instance, a career senior manager, myself, was instructed not to issue a valid order for payment of the additional royalty. Kerr-McGee knew the royalty was underpaid and the Director of MMS personally stopped the order from being issued.

Kerr-McGee has taken a stand that it owes no royalty on the deep-water leases that were issued without threshold values for royalty payments. These leases were issued by MMS in error by not including a threshold value for determining royalties due the Federal government. Some companies have renegotiated similar leases in an attempt to correct the error and bring an element of fairness to the American taxpayer. However, Kerr-McGee is aggressively pursuing the issue indicating that it will not pay any royalties to the American taxpayers on its deep-water leases with the missing threshold dollar value. However, Kerr-McGee will receive billions of dollars in revenue from the sales of oil and gas from these assets that belong to the American public.

I sincerely hope that this Congress will hold Kerr-McGee responsible for paying all royalties that it owes. I believe, and a jury of 12 American citizens agreed, that Kerr-McGee filed false royalty reports with MMS and did not pay its full royalty obligation. Further, I believe that behind closed doors in Washington D.C. the decision was made that Kerr-McGee would be let off the hook and not required to pay the royalties it owed. It was totally inappropriate for the Director of MMS, as a political appointee, to intervene and revoke my authority to issue a legal order to Kerr-McGee to pay the additional $10 million. Also, I hope Congress will review MMS' compliance program and encourage them to have a highly professional workforce and a truly professional audit program independent of political pressure.

This concludes my formal testimony. Thank you for the opportunity to appear here before this Subcommittee. I will be happy to answer any question you may have.

The Chairman. Thank you, Mr. Maxwell.

Mr. Gambrell?

Statement of Kevin L. Gambrell, Indian Land Working Group

Mr. Gambrell. Mr. Chairman and subcommittee Members, thank you for the opportunity to be here today. I see this as an opening to correct the course, encourage changes within MMS to act as a fiduciary and honor the treaty obligations with the tribes and individual Indians.

I have 16 years' experience in mineral industry work. My background is a Master of Science, Mineral Economics. I work as a practitioner for Alternative Dispute Resolution with the Morris Udall Foundation. I am connected with the Rocky Mountain Mineral Law Foundation. I have 280 hours of accounting auditing of law, and I have worked as a Navajo Nation mining financial analyst, as well as working for industry.

I came to the Federal government back in 1996, November 18, to run an office that managed Indian trust assets, oil and gas, on Navajo allotted lands. The reason the office was created was because the Federal government had breached the trust with Indian landowners, and many were not getting paid and were losing their homes, cars, livestock.

When I took over the office I was delegated the authorities of a BIA regional director, an MMS audit chief and a BLM field manager, so in my office I saw basically everything from the beginning of a lease, auctioning the leases, getting leases developed, to counseling leases and collected bonds. I have seen the whole process from end to end.

When I took over this organization, I ran into many areas of departmental resistance within the bureaus. Many of the bureaus were not willing to give up their authority, and I had to fight tooth and nail to get any authority within my office. I had to contact the deputy solicitor of Interior, as well as the deputy commissioner of Indian Affairs, to get action within the departmental agencies at the bureau level in the field.

At the end of the life of the pilot program, and it was a pilot project under Reinventing Government, Vice President Al Gore's pilot project, we were considered a success in 2002. During that time up until 2002, I had reported to a court and discussed what was happening within the Federal Indian Minerals Office (FIMO), explaining how we were in compliance with the consent decree that had orders to look at zero production, do audits, look at transportation, look at other issues, hire auditors, hire inspections, do everything from volumetrics to valuation.

In 2002 it was a success. We became departmentalized, and Gale Norton decided to extend this project to other locations in Indian Country. Also we had been nominated for the Hammer Award and other awards.

The things that have changed since I took the office and after 2002 when I stopped reporting to the court to me are a travesty. I saw a system called the compliance review system that did nothing in terms of really getting at the issues of valuation. I had an audit team that worked for seven years, and in five years of that seven years we always collected on back audit issues. Always. The compliance review process resulted in very little collection.

I also had situations where the system was shut down in 2001 in November, the MMS royalty system. Soon after, when that system came on line it was a new system that was created by Accenture, and it was to increase the compliance review process and make it a more contemporaneous, real-time audit process similar to an audit process. It is not quite. It is more of a compliance review.

When that system came on it failed miserably. I think over 50 percent of the royalty did not pay out in November 2001. In December 2001, Cobell litigation. Judge Lamberth issued an order to shut down the system, and that in a sense saved MMS from being exposed for the problems that were occurring within that system in November 2001.

The system did not come back on until 2002, April 2002. When that system came back on, many people had already lost their homes, cars, houses in Indian Country, but in addition to that the system still did not pay out correctly. MMS changed the error-checking system on volumetrics in order to let the data flow through the system and pay out.

That is just one example of the many problems within MMS and the compliance review system after I had stopped reporting to the court.

I thank you for hearing my testimony. If you have any questions, I would be glad to answer.

[The prepared statement of Mr. Gambrell follows:]

Statement of Kevin L. Gambrell for Indian Land Working Group

Mr. Chairman and subcommittee members, thank you for the opportunity to be here today. I see this as an opening to correct the course, encouraging change within Mineral Management Service (MMS) to act as a fiduciary and honor the treaty obligations with tribes and individual Indians.

I served and continue to serve Indian nations and the American taxpayer for sixteen years. I look back at my experience and recognize that it is one of the greatest joys of my life to see tribal people get what they are entitled to. However, I also recognize that MMS and other bureaus have forgotten their responsibility and now place industries desires and wants above their trust responsibility to maximize the benefit of oil and gas development to tribes and the American public.

I have a Bachelor's degree in International Trade and Relations and a Master's in Mineral Economics from the Colorado School of Mines. Further, I am a practitioner for Alternative Dispute Resolution with the Morris Udall Foundation and I am a whistleblower.

I started my career in oil and gas management with Navajo Nation as their Mining Financial Analyst and later went to work for the U.S. Government as the Director of the Federal Indian Minerals Office (FIMO).

The creation of FIMO was a result of Navajo individual Indian mineral owners, known as the Shí Shikéyah (roughly translates to, "This is My Land") Allottee Association, filing a lawsuit against the Department of the Interior in 1983 claiming that the federal government mismanaged their resources. After years of litigation, the U.S. District Court ordered Interior to establish FIMO.

FIMO was established in the early 1990s, however the employees mirrored the bureaus, in that they had few common goals, struggled over turf issues, handed off responsibility to other staff based on their bureau functions, and performed little to no asset management. The office failed to become the seamless, efficient and effective office the Department committed to, thus the Shí Shikéyah Allottee Association requested the court intervene.

In 1994, the Department established a National Performance Review Laboratory under Vice President Al Gore's reinventing government initiative. The laboratory is known as the FIMO Pilot project. Part of the initiative was to establish a FIMO Director that would navigate the staff from Bureau of Indian Affairs (BIA), Bureau of Land Management (BLM) and MMS and the mission. This would require delegated authorities to operate as a trustee for individual Indians in the Four Corners Area. The goal was to perform proper lease management, and accurately collect, disburse, and verify all royalties and volumes due from the severance of minerals. In addition, FIMO would be placed near the beneficiaries of these royalties so as to work with the beneficiaries directly.

In November of 1996, I was hired as the FIMO Pilot Director. Over the next six months, I changed the reporting relationship of the FIMO staff from the three bureaus to me, reported to a DC level Interior committee and acquired delegations of authority to act in the capacity of a BIA Area Office, a BLM District Office, and a MMS Compliance Division with regard to mineral issues.

Although FIMO was better equipped to act as the primary source for fulfilling the trust responsibility, the agencies maintained control, continued old practices and engaged in battles with me anytime I questioned their trust management practices. To counter this defiance, during the Pilot phase I found safety in reporting to the District Court, the allottee association, and the committee. I overcame most of the organizational resistance and turned around a negligent approach to managing mineral assets on Indian lands to providing services that met or exceeded the requirements of a Federal District Court consent decree. In one example, we annually collected 7 times the underreported royalty than MMS did for 20 years prior to my management. FIMO had the highest underpayment collection to audit cost in comparison to other MMS audit groups.

After a thorough evaluation of FIMO, it was considered a success and made permanent in October 2001. The Secretary of the Interior gave the green light to the committee to implement the FIMO concept throughout Indian country and I and my staff were nominated for two National Hammer Awards and numerous Spot Awards for excellent performance.

Once the pilot phase ended, the court reporting requirements stopped, and I reported to and took direction from line managers, I lost the shield and became challenged with managing the trust assets under the direction of management that made contrary decisions to maximizing the benefit to the Indian allottees.

Over time, I exhausted all efforts in attempting to protect the beneficiaries' interest against my superior's unethical decisions and decided to communicate with the Special Master of Cobell litigation who had the responsibility to ensure that information vital to the interests of individual Indians be safeguarded. I have experienced the following:

2000 Indian Gas Rule

In 2000, MMS changed the Gas Rules on Indian lands allowing companies to use index prices versus doing the full accounting in accordance with the lease requirements. As the Director of FIMO, I opted out of the pricing method because I did not believe I had the right to change lease terms without the consent of the Indian landowners and the method appeared to negatively impact royalties. When I voiced my concerns and decision, I was chastised by management and told that my decision was wrong and burdensome to industry.

When MMS implemented the regulation, the tribes and MMS made the decision for individual Indians, except Navajo allottees, to follow the new rule. At the start, the Navajo allottees were ahead of the index pricing by 25 to 45 cents per mcf every month. I reported this to management and they ignored the evidence that the New Rule potentially resulted in losses to the tribes and individual Indians. Soon after, DC management told me that I was not to attend the quarterly meetings with tribes and states.

They did nothing to inform other Indian groups that index pricing was producing negative results in the San Juan Basin and tribes had no way to benchmark the method without having gross value information. Although MMS had conflicting data, MMS failed to act and protect the interest of tribes and individual Indians. MMS management behavior violated the requirement to maximize revenues to the beneficiaries as stated in the regulations and law.

Undervaluation of trust assets, a violation of 64 C.F.R. § 43506, as the Minerals Management Service explicitly acknowledged with regard to valuation of gas production from Indian leases, it is responsible, ''[t]o ensure that Indian mineral lessors receive the maximum revenues from mineral resources on their lands consistent with the Secretary of the Interior's (Secretary) trust responsibility and lease terms.''; see also Federal Oil & Gas Royalty Management Act—1982—Title I, Section 101, Duties of the Secretary ''The Secretary shall establish a comprehensive inspection, collection and fiscal and production accounting and, auditing system to provide the capability to accurately determine oil and gas royalties, interest, fines, penalties, fees, deposits, and other payments owed and to collect and account for such amounts in a timely manner; and 30 U.S.C. § 1711 Comprehensive Accounting and Auditing.

2001 New Computer Compliance System and the Shut-down

In November of 2001, MMS shut down the old computer compliance system and turned on the new system created by the Bermuda based Accenture, previously Arthur Anderson. MMS did not parallel test the new system against the old system and the new system systematically failed halting royalty payments to more than half of the federal and Indian leases. In December, the problems were masked or decoyed by the court ordered shut-down resulting from Interior's failure to protect Indian data from internet hackers. Tribes and Indian individual stopped receiving royalties for almost five months and MMS management was able to save themselves from a congressional and public flogging.

The system came back on line at the end of April, 2002 and MMS still could not determine where to allocate monies placed in escrow. In desperation, MMS reduced their error controls and allowed erroneous data to pass through the system in order to get payments out. Only part the payment made it to the correct accounts. During the process, I made a request for the raw data to do my own reconciliation and management told me that their internet contractor had the data, but could not provide it because it was proprietary.

The claim by Interior that they were not able to pay tribes because they were off the internet was false. About five years before, Interior paid tribes using a manual system. During the entire shut-down, MMS and other bureaus blamed the Cobell litigation for the problem. Consequently, many Indian people lost their homes, automobiles and livestock.

In reality, this was a multi-million system not suitable for compliance, as required by Inspector General Act of 1978, as amended, (5 U.S.C. App. 3), which establishes as goals:

1. Promoting economy, efficiency, and effectiveness within the agency.
2. Preventing and detecting fraud, waste, and abuse in agency programs and operations.

3. Keeping the agency head and the Congress fully and currently informed of problems in agency programs and operations and of the necessity for and progress of corrective actions.

To this day, there are still monies sitting in escrow and the new system does not live-up to the compliance system that Accenture told MMS they would create. Accenture's contract was to end once the system was on-line, but their million dollar contract continues. Many tribes and state outright reject using the system for any serious analysis and resort back to the prudent and thorough audit approach. Lastly, how many times has the MMS Director and staff mislead congress on how well the compliance system works?

2000 to 2002 Closing Legacy Audits

In 2000, MMS made a decision to close all back-logged audits. To fully and diligently complete the audits, MMS would have to reverse their compliance review course and hire and train additional auditors with oil and gas accounting backgrounds. MMS decided not to change their course and made the deliberate decision to fast track all past audits with inappropriate and illegal valuation methods such as the ''bump method'' and other questionable practices. I blew the whistle on the actions management was directing me to do that I knew based on past audits would result in 1/8 the collections of underpaid royalties.

On January 31, 2003, my audit staff and I had a manager meeting in Denver regarding audit goals. We talked with MMS Managers about accounting requirements and we were directed to:

- bypass the negotiated settlement approval process,
- ignore third party verification,
- classify rudimentary reviews as ''Yellow Book Audits,''
- not document discussions with industry,
- use a compliance system that did not work, and
- perform duties and meet goals without adequate resources.

I raised concerns continually and provided evidence that their method would loose millions of dollars. MMS management ignored my arguments and used the fasttrack approach throughout Indian country.

While the ''Bump method'' was casually used in the past through a settlement negotiation process with Indian nations, it had never been lawful before 2000 as an alternative method without the required negotiated settlement process. The negotiated settlement process, as defined in Royalty Management Program, Audit Manual 2.0, 2.1.5, requires that if the company is unable to perform requirements under the Order to Perform, they may use the negotiated/settlement process with the approval of the Director of MMS and Assistant Secretary of Indian Affairs. MMS management willfully violated these rules and continued to implement their ''fast track'' procedures to meet their unreasonable goals.

As a result of my unwillingness to comply with such outrageous demands I feel that my superiors placed me in a situation with only three alternatives: (1) blatantly breach the trust of the beneficiaries, (2) act against management in insubordinate manner, or (3) leave. I chose to leave.

It is difficult to assess how many millions of dollars were lost because of this decision. Under any fiduciary system, this deception and corrupt practices would be considered malfeasance and negligence and somebody would go to jail. Even after I reported this problem to the Office of Special Council and the Office of the Inspector General, they did nothing to investigate or correct this problem.

2002 Trespass Issue

In 2002, I encountered a trespass issue where a company was operating a cancelled lease. The company produced for almost a year without reporting to MMS. The company received checks for production and cashed them monthly at a liquor store in Louisiana. We discovered the trespass when one of my inspectors visited the well site and found that it was producing and selling through the transporter. We contacted the transporter, told them that the producing company was trespassing and that we wanted all sales information. I contacted the trespassing company and told them we would collect 100% of the gross proceeds. They were upset, contacted their attorney and debated about how unfairly they were being treated. I told them that I would not tolerate this violation and would consider filling criminal charges against them.

The trespassing company contacted MMS management. Then an MMS Manager contacted me and told me that I could no longer talk with the company and scrutinized my evidence and my approach to the trespass. I asked management did they understand that they work for the American taxpayer. In order to circumvent ''Friends of the Company'' MMS management, I passed the responsibility to my auditor and forced the company to comply.

This case revealed how easily a company could circumvent the reporting system and produce and collect revenue without anyone's knowledge. This further emphasized the need for third party verification and field inspections. It also revealed that management considered industry complaints credible to the extent that they were willing to violate the beneficiary's interest, court orders and attack an MMS subordinate's position.

2003 Zero Production

After I left federal service, MMS and the Solicitor in Albuquerque, NM reversed my decision to collect additional value from companies violating lease terms of shutting in production without approval, known as ''Zero Production.'' The Solicitor wrongfully believed that liability did not transfer with change in lease ownership. The Solicitor decision is false based on private land oil and gas cases and general property law.

Before I left, I had already collected more than a million dollars and had about million dollars in cases still pending. Even though I was able to collect on past violations, the Solicitor ignored the results of my collections and arguments and reversed my decisions. I believe that MMS and the Solicitors lost more than a million dollars to Indian landowners. Their actions again violated court orders and MMS and the solicitors never looked anywhere else in Indian country to investigate ''zero production.''

Compliance Audit Tracking System

During 2000, MMS management discussed what systems they would continue with and remove. The Compliance Audit Tracking System (CATS) was marked to be removed and when I discussed what the system did—track all orders, issue letters and follow-up of compliance work over time—management stated that they had no idea that the system contained this data. This lead me to conclude that the MMS management was not consulting with auditors and were willing discard anything they did not understand or they intentionally wanted to remove historical information. Discarding this system was in fact a removal of the tool that helped the auditors track ''records of decision.'' In talking with auditors today, I have been told that MMS has a new system, but it does not cover tribes and states and I am concerned that past audit ''records of decision'' have not been included.

Conclusion

The problems of mismanagement of the public and Indian trust go far beyond the MMS royalty issues. I experienced the broad failures in protecting trust asset from BLM's expedited drilling and development approvals to BIA's right-of-way undervaluation. In today's environment, the actions of government executives represent an extension of industry, in which the federal managers fail to understand who they work for. Most federal executives have a company job waiting for them once the administration changes. The revolving door to industry has created a management team that is loyal to industry and the honest and diligent government worker is oppressed and pushed out of the way, thus violating and discounting the public trust. This is a travesty.

Recommendations

1. There is evidence that some oil and gas companies have not reported and paid royalties, therefore indicating that the system is still somewhat of an honor system. There must be third party verification through transporter and plant information.
2. Although MMS makes claims that regulation changes will benefit the American public, tribes and industry, it often benefits industry more. Once the regulation is implemented, there is often it is difficult to evaluate if the regulation changes actually benefit the tribes and the American public and MMS usually neglects this evaluation. Needless to say, industry tracks benefit to the penny for any regulation change. MMS needs to reevaluate the regulation changes and if they do no work, report this to the public and change course. The method of evaluating the regulations effectiveness must be emphasized and clear in public registrar at the outset of any proposed regulation change.
3. MMS's regulation changes that modify lease terms are a violation of contractual arrangements between lessees and lessors. MMS uses regulations to modify lease terms and inadvertently damages the interest of the landowner. MMS does not ask the landowner if they would sign a new agreement to clarify or improve lease terms, they simply make changes forcing the Indian landowners to comply. This violates

contractual and property law. MMS and other agencies must consult and obtain approval to change lease terms.

4. MMS needs to be accountable to the states, tribes, and federal government. As it is now, MMS reports and is accountable only to federal politically appointed executives. The federal government has a 50% stake in state leases and a 0% stake in Indian leases. As such, MMS should be guided, assessed, and managed by all government stake holders. A board of governors representing the states, tribes and federal government would do more to force MMS to be responsive to their concerns than the current UNILATERAL approach that is often politically manipulated with the industry stakeholders having the largest influence.
5. Although, MMS has limited resources, expediting settlements, compliance and collection at the loss of accountability makes them a ''simple paper processor'' with little to no concern for maximizing the benefit to the American public and enforcing lease compliance. MMS needs to change their objective from reporting false compliance information to maximizing revenue to the tribes and the public.
6. Everything MMS does must be made transparent and trackable. Any meetings with industry must be recorded and documented. Every action should be recorded in a system that can be queried by any state, tribal and MMS employee working on compliance. The IG should be able review these documented discussions and events at any time.
7. IG audits must go beyond the Yellow Book standards or Government Auditing Standards (GAS) in reviewing oil and gas audit work. Although, the GAS covers important issues such as transparency, accountability and peer reviews, it does not review the intricacies of valuation and volumetrics. For example, an IG auditor will look at the scope of work, internal controls such as signatures by managers, proper indexing and spreadsheets that add up in the total column, but they rarely review methodology of valuation and compliance with regard to court orders and other legal instruments. The IG must support positions that are well versed in mineral and energy accounting, as well energy law.
8. MMS must use the full extent of the lease terms to force compliance. MMS currently uses a penalty process to enforce royalty collection that is rarely collected and gets few if any results. Under royalty violation, I have used the cancellation clause to enforce the lease terms and companies have immediately taken action to comply.
9. With regard to whistle blowers, the staff within the Office of Special Counsel must be diligent and knowledgeable. They must follow-up on issues and hold management accountable. Retaliation laws need to be stronger, the investigative process needs to be thorough, and management needs to be accountable and punished when violating the public trust and employee rights.
10. MMS needs to restore the audit function. Determining underpayment is not an engineering calculation and requires an experienced oil and gas auditor that can look a vast array of data that is not only quantitative, but qualitative. The compliance review that MMS claims is an audit, is not and should be reported as only a review. MMS uses the misinformation and other false data to hype the compliance work that MMS claims they do annually. Many tribes and states have purposely removed themselves from relying on only compliance reviews as a realistic approach to lease compliance. These mineral assets belong to the public and Indians and once

exploited, are gone forever. Every penny owed to the public and tribes should be acquired with the most diligent and reasonable approach.

11. MMS must follow not only the laws and regulations, but also the court orders. This concludes my formal testimony. Thank you for the opportunity to appear here before this Subcommittee. I will be happy to answer any question you may have.

Responses to Questions Submitted for the Record by Mr. Kevin Gambrell

Would You Explain the ''Bump Method'' to the Committee?

The ''Bump Method'' is a fast track approach for the dual accounting requirement (Refer to MMS Audit Manual, Chapter 17, Section 17.5). This method usually results in fewer dollars collected for Indian allottees. In the following I will provide a written summary of the method and show the calculations at the end.

Dual Accounting Definition

Dual accounting is sometimes referred to as accounting for comparison and it is found in many, if not most Indian leases, both allotted and tribal. The purpose is to protect the royalty interest owner by comparing two values and using the higher of the two for royalty purposes. More specifically, a company acquiring a lease on Indian lands must pay gas royalty on a value which is the higher of the value of gas before processing less applicable allowances,(well head value), to processed less applicable allowances (the combined value of drip condensate, residue gas, and other gas plant products, less applicable allowances like transportation and processing). To determine the reasonableness of the company's reported value, the auditor must determine the gross revenue the company received for the gas and if it was based on the comparison for processed and unprocessed gas. This requires an audit of the company's records and systems, in which we look for purchasing contracts, allowance notifications, deductions, spreadsheets showing the comparison, and systematic problem due to internal controls and systems. Often, we go to third parties, such as gas plants and transportation companies to determine if deductions were reasonable. This requires an auditor with very specialized skill who understands the ''ins and outs'' of gas transactions.

''Bump Method'' Definition

An alternative approach to the dual accounting comparison is a method known as the ''Bump Method'' and also referred to as the alternative methodology and the fast track approach. The method was questionably approved under the Amendments to the Gas Valuation Regulations for Indian Leases, effective January 1, 2000. The reason I say questionably is that rule changes lease terms without the consent of the Indian lease holder. The concept behind the ''Bump Method'' is that the company will adjust the value of gas with regard to quality before processing to determine a value after processing, thereby bypassing the dual accounting comparison. More specifically, a company will take the quality of the gas and if it is greater than 1000 British Thermal Unit (BTU) the company will adjust the unprocessed value based on table of increments found in 30 CFR § 206.173. For example, company A has unprocessed gas quality of 1506 BTU. The company will refer to the table and find the adjustment under ''non-ownership in gas plant'' is in BTU range of 1501 to 1550, requiring an increment of .1600. If the company received $1.00 per thousand cubic feet of gas, the company would multiply $1.00 times the increment of .1600 plus 1. This means the company will pay royalty on $1.16. Theoretically, this approach captures the value of the additional products in the gas stream that increases the BTU quality. The caveat is that the auditor assumes the price is correctly reported and bulletin prices in the area are not manipulated. We know from the 1999 Qui Tam on crude oil, Benjamin Johnson versus Unocal, that companies collude and manipulate price indices. The companies essentially received value above the index price, but never paid royalty or taxes on the higher actual value they received. In addition, any audits pending for periods before 2000 had a much greater risk of falsified prices because of the market changes, deregulations, and vague and questionable practices of transportation, measurement and reporting.

The equation:

$$\text{Adjusted Value} = (\$1.00 \text{ X } .1600) + 1$$

And

$$\text{Adjusted Value} = \$1.16$$

The logic behind this method is that value for unprocessed gas adjusted with the increment will capture additional value in processed gas, therefore giving royalty the higher value. *The caveat is that the company's reported price and/or the index price may or may not be reasonable, and requires verification.*

This method was used in the past through a settlement negotiation process with Indian Tribes, but was not sanctioned as alternative method outside of the required negotiated settlement process. The negotiated settlement process, as defined in Royalty Management Program, Audit Manual 2.0 states in general that if the company is unable to perform requirements under the Order to Perform, they may use the negotiated/settlement process. The Manual than refers to the process which requires a settlement team, and once in final form, the practice has always been on Indian allotted land, to acquire approval of the Director of MMS, Deputy Commissioner of Indian Affairs, and the company.

<u>Bump Method (Alternative Methodology)</u>

A	*B*	*c*	*d*	*e* *b*d*	*F* *e-b*
Product	Reported Royalty	Quality (BTU)	Bump Factor	Adjusted w/Bump	Additional Due
Unprocessed Gas	$1,800	1100	1.0400	$1,872	$72

This method was approved as the ''alternative methodology'' on January 1, 2000 (Attachment 3) and 25 CFR 206.173. If used prior to Janaury 1, 2000 it required the ''settlement negotiation'' process because it was not legally available. Further supporting this position, it states in the Audit Manual Chapter 2, 2.2.5 ''...Always use applicable regulations for the period under audit...'' To understand why this is a fast-track approach, you must understand the ''Dual Accounting'' method approved prior to January 1, 2000. I will present it on the following page:

Processed Gas Method (Theoretical, Refer to Audit Manual, Chapter 17, 17.5.3 Used when company sell gas at wellhead, but processed down stream)

A	B	C	D (Mcf)	e	F	G	h (1000)	I (b*e)	j	K *	l **		
Products	Qty (Mcf)	Use (Mcf)	Shrinkage sum(b*e*f*g/h)	Gallons per Mcf	Gallon (Dth)	Plant Recovery	Quality (BTU)	Gal	Price/unit	Transport Deduct	Process Allow	Value	Computation
Gas Methane	1000	60	111				1.3		$2.00	$0.20		$1641.42	(b-c-d)*h*(j-k)
Liquid Ethane	1000			.2200	.065998	.73000		220	$0.22		.3333	$16.13	i*j*l
Liquid Propane	1000			.9000	.091065	.87000		900	$0.32		.3333	$96.00	i*j*l
Liquid Nor-Butane	1000			.2000	.102989	.90000		200	$0.39		.3333	$26.00	i*j*l
Liquid Iso-Butane	1000			.2000	.099022	.90000		200	$0.44		.3333	$29.33	i*j*l
Liquid Iso-Pentane	1000			.0050	.108790	.80000		5	$0.45		.3333	$0.75	i*j*l
Liquid Nor-Pentane	1000			.0050	.110102	.80000		5	$0.45		.3333	$0.75	i*j*l
Liquid Hexane	1000			.0350	.115060	.80000		35	$0.45		.3333	$5.25	i*j*l
											Total	$1815.63	

Product	Processed natural gas results in by-products in various liquid forms as well as methane gas. All products have value and the company must sum the values and compare it to the unprocessed gas value. The Indian owner gets the higher of the two.
Gallons per MCF	This is determined through a gas analysis report for each well.
Gallon (Dth)	Liquid products thermal heat value, decatherm.
Price/Unit	Bulletin or index prices for methane gas and gas products, reported monthly.
Use	Plants and field equipment require fuel which is taken out of the gas stream. This is allowed as usage and is not royalty bearing.
Plant Recovery	Every gas plant has unique efficiency or recovery factors. There will always be some loss, referred to as shrinkage, because plants cannot fully recover every molecule of gas and liquid.
Transport Deduct	When a company incurs transportation cost they are allowed to deduct it if they file the appropriate form. The reason we require forms filing is that a company may double dip. The company may report a net price with the transportation adjustments already taken and then the company would deduct the cost again, as if the price were not net. This issue often results in additional monies for the Indian beneficiary. Many companies fail to file forms and MMS has misplaced these forms. Refer to Audit Manual Chapter 19 and 30 CFR 206.104 and 206.105 (1996)
Process Allow	When a company incurs processing cost, they are allowed to deduct it if they file the appropriate form. As with transportation, the reason we require forms filing is that a company may double dip. The company may report a net price with the processing adjustments already taken and then the company would deduct the cost again, as if the price were not net. The reason we use .3333 is that processing allowances are limited to 66 2/3%. However, if cost are lower, the company must use actual. For example, the plant charges 50%, the company retains 50% for value versus 33%. Many companies fail to file forms. This issue often results in additional monies for the Indian beneficiary. Refer to Audit Manual Chapter 20 and 30 CFR 206.178 and 206.179 (1996)

Processed Gas Method (Actual)

I will not go through the calculations because it varies depending on the gas purchasing, transportation and processing contracts. I have several settlement statements between the purchaser and seller that show cost associated with processing and transportation and the values of methane and liquid by-products. **Often we find companies taking improper deductions, such as marketing cost, or gathering cost. Without attempting to obtain the data, we cannot determine what deductions were taken.**

Unprocessed Gas Method (Wellhead Value)

a	B	C (1000)	d	e	F	
Product	Quantity (MCF)	Quality (BTU)	Contract Price	Transport Deduct	Value	Computation
Unprocessed	1000	1.1	2.00	.20	$1,980	*b*c(d - e)*

This method is generally referred to as the wellhead approach or the BTU method.

In what ways has this impacted Indian Country?

The "Bump Method" has wide and negative impact on Indian Country. I have benchmarked the method against actual dual accounting and have found (8) eight times the amount of underpaid royalty.

Why did MMS exclude the Navajo Allottees?

I was delegated as the BIA Regional Director with authority over the Navajo Allotted leases and I opted out of the 2000 Gas Rule for two reasons: (1) the new method violated lease terms, and (2) the new method resulted in less recovery of underpaid royalties than the lease defined method of royalty valuation. I worked with my audit staff to determine this and tracked the value against the 2000 Gas Rule prices on a monthly basis. The Navajo Allottees realized a 25 to 45 cents per thousand cubic feet gain over MMS determined prices every month. The 25 to 45 cent gains adds up to hundreds of thousand dollars in realized royalty for Navajo allottees. This was before audit and audit would recover additional monies.

The "Bump Method" was applied to all other allottees or Individual Indians across Indian Country, except the Navajo that I excluded. MMS did not evaluate the benefits to this method.

Mr. Gambrell. You testified that while you were acting as Director of the pilot project of the Federal Indian Minerals Office you collected 7 times the underreported royalty than the MMS had for the 20 years before you took the position. Yet once this program became permanent you were stymied from aggressively going after underpaid royalties due.

Was this due to inertia and turf battles among the agencies at DOI? Do you see any way to change the attitude or throw some light onto it?

The past under performance of MMS was due to negligence and incompetence. Industry knows that MMS's upper management will fold almost anytime they are threatened with litigation. MMS has and has had an incestuous relationship with industry. Most managers over MMS walk through the revolving door to industry and back again. These executive are inherently serving their own self-interest above and beyond Indian and Public trust. The only way to change this attitude is to:

1. Limit the ability for Senior Executive Service managers to work for Industry in the area that they had influence over.
2. Limit the ability for politically appointed officials to recruit Executives from industry.
3. Use the existing ethic laws on the "Appearance of Conflict of Interest" more aggressively.

4. Change MMS's oversight and reporting requirement from a Federal Executive to a Board of Directors composed of Federal Executives, State Audit Managers, and Indian Audit Managers. The State and Indian Nations have a higher vested interest in assuring that their royalty interests are collected 100%.

Mr. Gambrell. You refer in your testimony to MMS problems with the computer system purchased by a Bermuda company which allowed erroneous data to pass through the system and eventually cause the shut down of the system.

Is it true that this shut down was falsely blamed on the Cobell litigation? Were Indian lease holders directly hurt by the shut down? How was the system fixed to ensure the correct data was in the system?

Throughout Indian Country, MMS, BIA and OST management blamed the shut down on the Cobell litigation, rarely referring to the hacker penetrable conditions of their network, which enabled any hacker on the street to access Indian data and manipulate it. During the same time, the MMS system was upgraded and it failed to account for collections and disbursements of Indian Royalties. We noticed the problems with the MMS system one month before the shut-down ordered by Judge Lambreth of the Cobell litigation. In addition, the Department of the Interior was able to pay Indian lease holder by shipping digital tapes from Denver, Colorado to Anadarko, Oklahoma. After the shut-down, the agencies would not exercise this method, for it would expose MMS system failures not related to Cobell.

Indian lease holders were hurt by the shut-down. Many lost automobiles, homes, live-stock and their credit was severely damaged.

The system was never fixed, but instead MMS changed the filtering of data to be less stringent, thus allowing garbage data and reporting to pass through the system without any scrutiny.

Mr. Gambrell. When you went to the Special Trustee for Indian Trust Funds with your concerns about the manner in which the trust responsibility was being carried out, what kind of a response did you receive?

When I speak of talking with the Special Trustee, I am referring to the Indian Minerals Steering Committee which includes Executives from BIA, BLM, MMS, Solicitors Office and OST (all have primary trust responsibility). When I would refer to issues, often the Executives would battle over turf issues, do as little as required by law, and interpret law to reduce their agency legal risk even if it meant damaging the benefit to the Indian individuals or tribes.

I also met with the Special Master under the Cobell Litigation, who had responsibility to oversee the retention of records and information. He was deeply concerned and found that what I had reported as egregious acts were in fact true. He reviewed the issue concerning audit records held in MMS Dallas office and found them to be missing, incomplete, and misfiled. He looked into the right-of-way issues and found the appraisal records were missing and destroyed. During his investigations he was obstructed by the Department Interior Executives and Secretary Interior's hired legal counsel.

Mr. Gambrell. You testified about a 2002 trespass issue which you found a company stealing the resource of an Indian lease and tried to bring the culprit to justice only to be

stymied by MMS. Do you know if the Indian allottee was ever paid for his stolen trust asset?

The allottees were paid because I delegated the resolution and tracking of this issue to my auditors. I was prevented from meeting with the company and so I delegated my audit staff to act on my behalf in dealing with the company.

Mr. Gambrell. Would you please speak to what is considered the revolving door scenario at the Department which leads to employees more loyal to industry then to those for whom they are trustee? How prevalent a problem is this and can it be stopped?

I will provide names of those walking through the revolving door. MMS Director Cynthia Quarterman, MMS Director Jonny Burton, Deputy Secretary of Interior Griles, Secretary of Interior Gale Norton, and so on...

The Executive Branch should:

1. Limit the ability for Senior Executive Service managers to work for Industry in the area that they had influence over.
2. Limit the ability for politically appointed officials to recruit Executives from industry.
3. Use the existing ethic laws on the ‘‘Appearance of Conflict of Interest’’ more aggressively.
4. Change the oversight and reporting requirements of MMS from a Federal Executive to a Board of Directors composed of Federal Executives, State Audit Managers, and Indian Audit Managers. The State and Indian Nations have a higher vested interest in assuring that their royalty interests are collected 100% and will not be as likely to be politically manipulated by an Executives political party affiliation.

Questions for Kevin Gambrell from Minority

1. You written statements is replete with accusations of unethical and illegal behavior by MMS, the Department of the Interior, and oil and gas companies. Are you and auditor or an attorney?

I was delegated with the trust authority to audit, inspect and enforce, and lease Indian allotted leases. I supervised and managed auditors, inspectors and lease administrators. I issued subpoenas, orders of non-compliance, orders to pay, negotiated settlements, and made all trust decisions with regard to oil and gas leasing on Indian allotted leases. My title equivalent was a Bureau of Indian Affairs Regional Director, Minerals Management Service Chief Compliance Officer, and a Bureau of Land Management Field Office Manager. My audit, inspection and enforcement and leasing authorities were delegated to me by the MMS Director, BLM Director, and BIA Deputy Commissioner. I had the authority to enforce all laws and regulations that pertained to the Code of Federal Regulations 25, 30 and 43 with regard to mineral development on Indian Allotted lands.

The question as to auditor and attorney is irrelevant with regard to upper management positions within the MMS, BLM and BIA. The Director of MMS, Johnnie Burton has a Bachelors degree in Education, while I have a Master of Science in Mineral Economics and have taken more than 280 hours of GAAP, FASB, and valuation/auditing training. In addition, I was required to take auditing continuation credits annually, where as the current Director does not take auditing continuation courses.

Did the auditors in your office or solicitors with the Department agree with you?

The auditors in my office fully agreed with me and the solicitors, although often they lacked any significant experience, usually agreed with me. The solicitors' agreement depended on the issue. I only considered the Solicitor's advice as a legal opinion that I would consider in making my decision, but would not always go with.

Did the Inspector General of Interior or Office of Special Counsel agree with you when you brought it to their attention?

The two organizations are not staffed with professionals to address valuation, trust and other technical issues. The IG often reviews only the surface issues that pertain to ''Yellow Book'' standards, but do not consider actual methods of determining value and bench marking. The OSC did not comment on my issues and did not follow-up with questions or concerns.

2. You say MMS willingly ''violated'' court orders in the 2002 trespass case and the 2003 Zero Production case (page 5 and 6). Which Court's Order were violated, and what did the Court do about it?

Shii Shi Keyah Association v. Hodel, No. Civ-84-1622M D.N.M. 1984)Hon. E.L. Mechem). A class action on behalf of all Navajo allottees with oil and gas leases on their lands alleging that the Secretary of Interior was not in compliance with the Federal Oil and Gas Royalty Management Act of 1982. A consent decree was entered in March of 1989 requiring numerous changes on the part of the Secretary to bring the Secretary into compliance with FOGRMA. This was after several key summary judgment rulings in Plaintiff's favor on interpretations of key provision of FOGRMA. An award was also entered for the Plaintiff under the Equal Access to Justice Act which was, at that time, one of the largest such awards yet to be entered. The court retained superintendent jurisdiction over implementation of the consent decree for five years.

The Department of Interior reporting requirement ended in 2001. Consequently, the court is not aware of these violations, but will be notified through additional lawsuits.

I also met with the Special Master under the Cobell Litigation, who had responsibility to oversee the retention of records and information. He was deeply concerned with violations of records retention and found that what I had reported as egregious acts were in fact true. He reviewed the issue concerning audit records held in MMS Dallas office and found them to be missing, incomplete, and misfiled. He looked into the right-of-way issues and found the appraisal records were missing and destroyed. During his investigations he was obstructed by the Department Interior Executives and Secretary Interior's hired legal counsel.

The Chairman. Thank you.

Ms. Alexander?

STATEMENT OF RYAN ALEXANDER, PRESIDENT, TAXPAYERS FOR COMMON SENSE

Ms. Alexander. Good morning, Chairman Rahall, Ranking Member Flake and Representative Costa. Thank you for the opportunity to testify today.

Taxpayers for Common Sense is a national, nonpartisan budget watchdog organization. We believe in transparency, competitive and clean contracting and accurate and independent auditing. In short, we believe that taxpayers have a right to demand excellence in accountability from our government.

For more than a decade, TCS has worked actively to ensure that taxpayers receive a fair return on the resources extracted from Federal lands and waters. In recent years, numerous management failures at the Minerals Management Service have cost taxpayers billions of dollars in waste and lost revenue.

TCS urges the Committee to reform the revenue collection process, to improve contracting practices and to increase accounting accuracy at MMS. We also urge the Committee to hold the oil and gas industry accountable for accurate reporting of minerals extracted from Federal lands and to eliminate royalty relief programs. We look forward to working with the Committee to effect these changes.

My focus today will be on three areas of primary concern to Taxpayers for Common Sense: The risks presented by the reliance on compliance review and industry self-reporting, the risk to taxpayers the Royalty-in-Kind program presents and the failure of MMS to remedy known problems.

It is the responsibility of MMS to ensure fair calculation, collection and distribution of royalties on behalf of the American taxpayer. In decades past, the MMS Auditing and Compliance Division collected over $100 million annually through the audit process. In recent years, this amount has declined to less than half that number.

In the last decade, MMS began transitioning from a traditional audit process to a new automated royalty verification process known as compliance review. This system relies substantially on self-reported data from the oil and gas industry.

A recent report from the Inspector General concluded that the compliance review process may not detect underreporting and underpayment of royalties, particularly because anomalies detected in the compliance review process rarely trigger a traditional audit. The combination of self-reporting and superficial data reviews provides companies with an incentive to underreport and underpay royalties owed.

The Royalty-in-Kind program usually puts the Federal government in the position of marketing oil and gas directly. We have concerns about putting the government in this position. Industry has proved itself very capable of bringing oil and gas to the marketplace. Putting the government in that role is potentially costly and inefficient.

Reports from the Inspector General raise questions about the ability of MMS to track royalties collected, to track the volume of production on Federal lands, and earlier this morning we heard Mr. Gaffigan raise questions about MMS' ability to determine whether or not the sales from royalty-in-kind payments equal or exceed cash royalty payments as required by statute.

Given these concerns, we have little confidence that MMS is equipped to get the best deal for taxpayers through direct sales. At the very least, the royalty-in-kind system should be thoroughly evaluated, and if found not to benefit the taxpayer we believe the program should be scrapped.

As the Committee well knows, the error of omitting price threshold language from leases executed in 1998 and 1999 has already cost the taxpayer over $1 billion. Recent reports suggest that Director Johnnie Burton was aware of this error in 2004, and it was brought to wide public attention by the New York Times expose` last year, and yet MMS has only recently begun to remedy this problem.

We are pleased to see that they have started to remedy this problem, but every single delay costs the taxpayers more money. Moreover, the current leadership at MMS has shown little appetite for pursuing underpayments discovered by their own staff. Worse, employees who have attempted to remedy underpayment collection have been dismissed.

It is clear that several actions at MMS must occur to remedy the current situation. Compliance review based on self-reported data cannot be relied upon to ensure adequate collection of royalty revenue. Steps must be taken to ensure independent audits occur and royalty underpayments cease. The system has to be reformed so it is more transparent and can easily account for royalty payments.

For example, when Assistant Secretary Allred was testifying earlier he couldn't give the numbers that you wanted. Those numbers should be very easily accessible. Furthermore, the system should be publicly accessible via the internet.

It is the Federal government's responsibility to protect the taxpayers' resources and ensure that they are adequately compensated for their sales. It is clear that the agency responsible for the taxpayer protection is in need of an accountability overhaul.

We are pleased to see such rigorous oversight by Congress. The absence of the checks and balances inherent in the oversight process invariably leads to problems, particularly in agencies that by the nature of their missions have close ties to the industries they regulate.

We are pleased the Committee has begun to address these issues, and we look forward to working with you to see the Minerals Management Service reformed.

[The prepared statement of Ms. Alexander follows:]

Statement of Ryan Alexander, President, Taxpayers for Common Sense

Good morning Chairman Rahall and members of the committee. Thank you for the opportunity to testify today. My name is Ryan Alexander and I am the President of Taxpayers for Common Sense (TCS), a national, non-partisan budget watchdog organization. The mission of Taxpayers for Common Sense is to fight wasteful government spending and subsidies to achieve an efficient and responsible government that lives within its means. We believe in competitive and clean contracting —from the Iraq war, to Katrina, to DOD procurement, to MMS contracts. We believe in transparency: taxpayers should be able to easily see how their tax dollars are spent, whether in the $460 billion defense budget or $300 million MMS budget. We believe in accurate and independent auditing. In short, we believe that taxpayers have a right to demand excellence and accountability from our government.

For more than a decade, TCS has actively worked to ensure that taxpayers receive a fair return on minerals and resources extracted from federal lands and waters. The mismanagement at the Mineral Management Service (MMS) offends all our core values and in the absence of corrective action will continue to waste tax dollars. TCS is committed to reforming our revenue collection process, ensuring fair contracting, and increasing accounting accuracy at MMS. TCS is also committed to holding the oil and gas industry accountable for fair and accurate reporting of minerals extracted from federal lands and supporting efforts to eliminate royalty relief provisions. We will continue to actively pursue each of these goals and look forward to working with the committee on other efforts to achieve these ends.

In addition to the mismanagement and enforcement problems at MMS, we believe there are structural problems with the current royalty system that subsidize the oil and gas industry at the expense of the taxpayer.

As you know, oil and gas companies that drill on federal and Indian lands or offshore pay royalties for the oil, gas and some other minerals they remove. Generally, this payment is a percentage of the total value of the oil or gas extracted. It is the responsibility of MMS to ensure fair collection, calculation and distribution of royalties on behalf of the American taxpayer. The collection of royalties is a significant source of revenue for the federal government: In Fiscal Year 2006, the Minerals Management Service reported more than $10 billion in royalty revenue.

Royalty Relief

With the oil and gas industry continuing to experience record profits, there is little need for taxpayers to continue to subsidize it.

Given the current fiscal climate, we commend the House for recognizing the need to reel in royalty relief provisions. Earlier this Congress, the House passed legislation requiring the repeal of royalty relief provisions included in the Energy Policy Act of 2005. We also applaud the MMS for proposing the repeal of sections 344 and 345 of the Energy Policy Act in their FY08 budget request. Taxpayers for Common Sense opposed the inclusion of these provisions in the Energy Policy Act and looks forward to working with Congress and MMS to see these sections repealed.

Royalty-in-Kind Program

Another area which Taxpayers for Common Sense fears is ripe for abuse is the Royalty-In-Kind program. From our standpoint, ''in kind'' contributions across government programs almost always end up being a bad deal for taxpayers. We saw a red flag when MMS began pursing an expansion of their in-kind program in the mid-1990s. The Royalty-In-Kind program allows oil and gas companies to pay their royalty dues in the form of oil or gas instead of cash. This forces the federal government to market the oil and gas themselves. The burden of marketing and selling oil and gas complicates government bureaucracy and leads to a lack of transparency.

It may be true that the Royalty-In-Kind program makes it easier for MMS and the industry to calculate the royalties that are due because they need only determine a percentage of the amount of oil produced and do not need to be concerned with the sale price. But the benefit for the government ends there. In effect, the process adds layers of complication and inefficiency by requiring the federal government to resell oil and gas. Involving the government in the sale of oil can easily lead to abuse. Given the current track record of MMS, we have little faith that this system can operate efficiently and for the benefit of taxpayers.

Auditing and Compliance

To ensure the adequate collection of royalties, MMS has an auditing and compliance division whose goal is to oversee leases and complete audits. In decades past, this division collected over $100 million annually through the audit process. However, in recent years this amount has significantly declined to less than half of that number. In fact, a more than twenty percent decrease in the number of audits was reported in the last five years. Not only has the collection of revenues dropped dramatically in recent years, MMS has clearly shown less commitment to this division, as demonstrated by shrinking budgets and significant cutbacks in staffing.

In the last decade, MMS began transitioning from a traditional audit process to a new, automated royalty verification process, known as compliance review. This shift has not been cost-effective and is an important contributing factor in the drop in revenues collected by MMS. Further, relying on self-reported data from the oil and gas industry is not an accurate way to monitor royalty collection.

Proponents of compliance review correctly point out that the process allows MMS to check data pertaining to more transactions than the traditional audit process; however, the superficial review does not allow for an in-depth analysis or encourage improved accounting procedures. As the Department of Interior Inspector General Earl Devaney testified before the committee in February, the compliance review process does not provide the same level of detail or accuracy a traditional audit provides.

The IG audit report released in December 2006 detailed many weaknesses in this program. The report highlighted MMS's inability to access accurate and complete information on the program and the inability to use it for daily management and reporting purposes. Further, the current system does not provide states, tribes and Congress with accurate information on the Compliance and Asset Management Program.

The report went on to conclude that MMS could not establish the true cost and benefit of compliance reviews and audits. When considering the impact on federal taxpayers, one of the most egregious findings of the IG report was that anomalies rarely lead to a full audit. The report concluded that "MMS may not detect underpaid royalties."

Additionally, the fact that the data relied on for this process is self-reported by the companies should be of grave concern. The combination of self-reporting and superficial data reviews provides companies with an incentive to under-report and under-pay royalties owed.

As demonstrated in the case brought forth by our fellow witness, Mr. Bobby Maxwell, as well as other auditors at the agency, a negligent MMS appears to be serving the interests of the oil and gas industry over those of the taxpayer. In a glaring example of mismanagement within the agency, these auditors were prohibited by the MMS from collecting gross underpayments of royalties they had uncovered in their investigations.

Contracting and 1998 and 1999 Leases

Perhaps the best-known example of mismanagement at the MMS is the errors made in the leasing contracts of 1998 and 1999. In 1995, Congress passed the Outer Continental Shelf Deep Water Royalty Relief Act which awarded the oil and gas industry a waiver of royalty payments for leases issued from 1996-2000. These leases were all intended to include price thresholds that would trigger the collection of royalties when the price of oil reached above $36/barrel.

A little more than a year ago it came to light that a gross error had occurred in more than 1,000 leases issued in 1998 and 1999. Contracts had omitted the price threshold language, unlike those for leases issued in 1996, 1997 and 2000. When this error was uncovered in a New York Times expose, a series of congressionally driven investigations determined it was merely a clerical error. This clerical error has already cost taxpayers at least $1 billion in lost revenue.

Adding insult to injury, Johnnie Burton, Director of MMS, was made aware of the error as early as 2004, despite congressional testimony she had given late last year to the contrary. The information was uncovered by the Interior Inspector General and documented in a series of emails sent to Ms. Burton.

While the original omission of the price threshold language was obviously a very serious error, MMS's failure to devise and implement a fair remedy in the nearly three years the agency has been aware of the problem is emblematic of the lack of accountability and culture of mismanagement at MMS.

On the subject of price thresholds, we would like to call one additional matter to the Committee's request. MMS finalized the "Shallow Water, Deep Natural Gas" rule in 2004. The rule is designed to spur development of natural gas far below ground in shallow waters. Unlike the 1998 and 1999 leases, this rule does include a price threshold. Unfortunately, MMS set the price threshold at the sky high level of $9.34 per thousand cubic feet of natural gas. The threshold is indexed to inflation and rose to $9.91 for 2006. MMS data show that this threshold is so high that companies would have avoided royalties even in 2005 and 2006, in a time of record high prices following the Gulf Coast hurricanes. A threshold this high is no better than no threshold at all.

We would also note that the threshold increased dramatically as the shallow water deep gas rule moved forward—from $5.00 in the proposed rule to $9.34 in the final version. The result will be billions in foregone revenues for the federal taxpayer. We encourage the Committee to look into this matter in greater depth.

Culture of Mismanagement at the MMS

Federal taxpayers continue to bear the burden of these multi-billion dollar errors. The problem will only be compounded in the coming years. Director Burton has shown little initiative to remedy the problems within the agency. In addition to failing to correct the missing lease language when it was first brought to her attention, several employees who have attempted to remedy the under-collection of royalties have been dismissed on her watch.

The Department of Interior estimates in the next five years that energy companies will likely extract $65 billion in oil and gas from federal lands and pay little or no royalties for it. This will cost taxpayers $7-$9 billion in lost revenue. The problem will only escalate as more oil comes online. By 2011, the Department of Interior estimates that royalty-free oil will quadruple and natural gas will see a 50% increase. Taxpayers cannot afford to have a grossly mismanaged agency overseeing this important source of revenue.

Remedies and Solutions

It is clear that several actions at the MMS must occur to remedy the current situation. Senior employees must be held accountable for their actions and committed to the mission of the agency, not the pocketbooks of Big Oil. Too many examples of close connections with the oil industry have surfaced to ignore this problem. We encourage the committee to continue rigorous oversight to ensure MMS is operating in the interest of federal taxpayers.

Furthermore, compliance review cannot be relied upon to ensure adequate collection of royalty revenues. Steps must be taken to ensure independent audits occur and royalty underpayments cease. The current system heavily relies on self-reporting, which can only lead to abuse. The system has to be reformed so that it is more transparent and can easily account for royalty payments. Furthermore this system needs to be publicly accessible via the Internet.

Past errors must also be corrected. Contracts that omitted the price threshold language must be renegotiated. We applaud Congress for beginning to take steps in this direction. It is clear in testimony provided by several of the oil companies involved with leases that the industry was aware of the error and was also fully aware of Congress's intent to keep the price thresholds in the contracts. Contracts are renegotiated all the time, and this situation must be addressed or taxpayers stand to lose billions more. Gross negligence on the part of government employees is an unacceptable reason to allow the oil and gas companies to exploit congressional intent and avoid the dues that are rightfully owed to taxpayers. These oversight hearings will help reveal to the general public any companies that refuse to pay. As we have already mentioned, it is outrageous to imagine giveaways to oil and gas companies while they are experiencing such enormous profits.

It is the federal government's responsibility to protect taxpayers' resources and ensure they are adequately compensated for their sale. It is clear the agency responsible for this taxpayer protection is in need of an accountability overhaul.

MMS's Royalty-In-Kind system has fundamental flaws that make it hard for taxpayers to be sure they are getting their money's worth from their resources. Under the best conditions, this type of system would be prone to abuse, particularly at an agency as flawed as MMS. At the very least the Royalty-In-Kind system should be thoroughly evaluated, and, if not found to benefit the taxpayer, scrapped.

The oil and gas industry runs on a boom and bust cycle. While seductive, the offer of royalty relief to stimulate production can skew the marketplace and have longterm unintended consequences of diminished returns for taxpayers. We urge Congress to be very judicious before pursuing royalty relief in the future.

Again, we are pleased to see such rigorous oversight by this Congress. The absence of energetic oversight or the checks and balances inherent in the oversight process invariably leads to problems, particularly in agencies that by the very nature of their missions have close ties with the industries they regulate. We are pleased the committee has begun to address this issue and look forward to working more to see this embattled agency reformed.

The Chairman. The Chair thanks the panel for your testimony. Let me begin my questions with Mr. Maxwell if I might. I do thank you for traveling as far as you have to stand up and be counted. The root of your testimony, it seems to me, is that since the year 2000 a dramatic different philosophy has taken root at MMS, and it appears that that philosophy is do not bother the oil companies. I believe you put it that way in your testimony.

It just boggles the mind. It is incredible. It is incredible that such a philosophy would take hold when it comes to the American people's lands. It is something the American people must know, and they are deserving of a true record.

Government auditors are told not to bother those whom they are lawfully entrusted to audit. If this system was in place at the IRS, I really fear that this country would be broke today.

My question is what possible gain do you guesstimate, Mr. Maxwell, that could be had by those who promoted this philosophy? Is there a benefit that you see they might derive from such a philosophy of just hands off the oil companies?

Mr. Maxwell. Sir, I don't understand the political ramifications or how these things change, but the whole culture has changed.

Changed dramatically.

I believe in the years preceding I was very aggressive in pursuing the oil companies. If we put an order out and we are pursuing something, I don't care if the oil companies sue us. If we are right and we need to establish a principle or enforce a regulation or a law or to find out if something is, let us litigate it. There are hundreds of millions of dollars or billions, and I think to be scared that we are going to be sued by an oil company is ridiculous.

The culture changed, and it changed in such a broad scale it was almost unbelievable because there were statements made. Don't bother the oil companies. Let us go to this new compliance system. We do not want you to be getting records like you did before, you know, getting massive amounts of records.

So it changed drastically, and a lot of the people we were hiring and putting in positions for compliance reviews had different backgrounds. The basic requirement used to be to be an auditor or an accountant you had 24 hours in accounting, which could include business law, statistics, et cetera.

We quit pursuing that type of background, and we would move different people in maybe from administrative areas or different areas. I am not opposed to doing that, but I think the basic requirement should have been if you move into those positions go get your 24 hours of college level courses because when we audit oil and gas companies they have some of the best and brightest, and any gray area they are not going to pay royalties on if they don't have to.

I don't blame them. If I was on their side, any gray areas I would interpret to my benefit. However, you need people that can realize what the accounting records are, different methods of depreciation, transportation, and to be able to apply the regulations, laws and lease terms. You have to have people that can think, can work diligently and would be on site at the oil companies.

And so when we started pulling away from that back to a compliance review to where people would work in the office and try to review without going on site to the companies, without talking to the marketing personnel, the field personnel, you really lose a lot of expertise in that culture.

I think that is why we saw the huge drop in royalty collections. I don't think it was because the oil companies just turned honest one year. I think it is because we stopped auditing.

I might add, when we have talked about compliance reviews there is a definite place for compliance reviews, but I think you don't start with a compliance review and then see what you need to audit. I think you start with the detailed audits, and then you back off into what compliance reviews are sufficient to monitor on an ongoing basis.

The Chairman. Was there any time that you conducted an audit and you found that money was not owed?

Mr. Maxwell. I don't believe I ever had an audit that we didn't collect anything.

The Chairman. The answer is no?

Mr. Maxwell. The answer is no, sir. That is right.

The Chairman. Mr. Gambrell, would you answer the same question?

Mr. Gambrell. Yes. No.

The Chairman. The answer is no as well.

Let me explore with you just a bit more, Mr. Maxwell. Was this a philosophy that you would say started at the grassroots within the MMS and percolated up, or was it something that started at the top and flowed downward?

Mr. Maxwell. Definitely started at the top and flowed down. We were directed that we would support this new reengineered method, the compliance. We would. We were directed at senior manager meetings that dissent would not be tolerated. They said we are making the change. We are making the transition, and it is your job to go out and sell it and enforce it.

The Chairman. So it was a trickle down philosophy then?

Mr. Maxwell. Yes, sir.

The Chairman. Mr. Flake?

Mr. Flake. Thank you.

Mr. Maxwell, I am interested in you were at MMS until? When did you leave?

Mr. Maxwell. 2005 the last time.

Mr. Flake. 2005?

Mr. Maxwell. Yes, sir.

Mr. Flake. What is the appropriate balance where somebody works and gets official information and then decides to go out on their own and file under the False Claims Act? Why wouldn't more people do that using information that way, and where is the proper balance between recouping money for the taxpayer or somebody's private interest doing that?

Mr. Maxwell. I think that is a great question because I looked at that, thought about it and struggled with it also.

In your employment situation, our responsibility is to the American taxpayer and also to MMS, our employer. I think that our job is to bill and the underpayments, to pursue them, to litigate them, through the MMS or the government when it will do it.

I think the False Claims Act only seems applicable to me when the agency refuses to do its job, and I think there could be three reasons that happens. One could just be corruption. Two could be just ignorance. You could have people that are incompetent. Three, you could have not enough personnel to do the job.

I think there has to be extraordinary circumstances for an employee to file a false claims on their own.

Mr. Flake. And you felt that that test was met here in this case?

Mr. Maxwell. I do, sir, for a couple of reasons. Number one is the director was aware of the issue, the Office of the Solicitor. I discussed it with the Office of Enforcement, so I believe it was well known. I was told the order would not go out.

Mr. Flake. Is there a protocol? Is there an order that you have to go through, and did you follow that?

Mr. Maxwell. I believe I did. I discussed it with the Office of Enforcement, and at one time in 2003 I had an interview with the Inspector General on another matter, which was the matter of not being able to audit royalty-in-kind.

It wasn't set up for this specific reason of this Kerr-McGee because I didn't know for sure at that time, but on the other meeting that was set up, the meeting never took place.

Mr. Flake. MMS disputes some of these meetings or the followup.

Do you have any comment on that?

Mr. Maxwell. I don't doubt that. No, I have no comment, but it doesn't surprise me.

Mr. Flake. Ms. Alexander, what is your feeling on that balance? Should a former employee be able to file under the False Claims Act using information that was gathered when the person was an employee? Where do you strike the best balance for the taxpayer in this regard?

Ms. Alexander. You know, I am not an expert in the False Claims Act so I don't know the legal and technical requirements under the Act. I think from our perspective we are happy that anybody is pursuing royalties on behalf of the taxpayers, and it would obviously be the best situation if there weren't a need, if there was no need for a False Claims Act.

I think it is difficult for me to really testify in general about a balance without knowing more about the law.

Mr. Flake. In your opinion, has MMS straightened up over the past year or so?

Ms. Alexander. It is hard to tell, but I certainly liked some of what I heard this morning. It sounds like they are making efforts to be more aggressive about collecting royalties, but, given their track record, you know, we are going to look at that with a great deal of skepticism.

Mr. Flake. Under the False Claims Act, Mr. Maxwell, how much do you stand to gain if the current numbers carry forward?

Mr. Maxwell. We currently have no judgment entered in the record by the judge so we don't have any dollars, but when the judgment is and if we eventually win, 100 percent of that money goes to the Federal government. Then we deal with the Federal government, but the percentage could range from 25 to 30 percent of the gross amount.

Mr. Flake. With regard, going back again to MMS and your time there, are there the proper protocols in place right now for an employee to say I have exhausted my remedy within the organization before it goes out, or do we need to change those?

Mr. Maxwell. I have been gone for several years and so I am not sure, but I think it should be well in place and very well communicated, but I really don't know what they have right now at this point.

Mr. Flake. You have been gone since 2005?

Mr. Maxwell. Yes. I have been gone for over two years.

Mr. Flake. All right. Thank you, Mr. Chairman.

The Chairman. Ms. Alexander, let me ask you. Do you agree with the statement I made that if the IRS were run like this agency was that our country would be broke?

Ms. Alexander. It seems like a kind of broad statement. I don't have enough knowledge, but I will agree with it.

The Chairman. Mr. Gambrell, let me ask you. Mr. Lester will state in a few moments in his testimony that Federal offshore mineral leasing laws do not allow tribes to prohibit oil companies from reducing payments by claiming overpayment in previous years.

You claim, and I quote, ''Unilateral adjustments are often made without the tribe's knowledge and lack review or oversight by MMS.'' If that is accurate, this seems quite egregious to me, and I would ask you if you would concur with Mr. Lester.

Mr. Gambrell. I am not quite sure what is meant by the offshore as an offset to Indian leases. I don't understand what that means.

The Chairman. Is there no statute of limitations, I guess is my question, for companies to claim overpayment?

Mr. Gambrell. My understanding is there is a statute of limitations to claim certain adjustments.

I also am aware that there are adjustments made in very small amounts that add up to the millions of lines of adjustments that go back quite some time to the 1990s, maybe even into the 1980s. I have looked at royalty reports where I have seen adjustments back in the 1980s.

The Chairman. Are companies required to provide proof of overpayment to the Department of Interior?

Mr. Gambrell. Yes, if they want to recoup. Well, within my office —I can't speak for the entire MMS, but within my office—we required some type of documentation to recoup.

The Chairman. And who would have authority over these issues?

Mr. Gambrell. Us.

The Chairman. And do you have any recourse?

Mr. Gambrell. If the company claimed an overpayment without justification, we could collect. We might collect a bond. We may issue a notice of noncompliance. We may cancel the lease.

The Chairman. OK. That ends my questions.

Do you have any further questions, Mr. Flake?

Mr. Flake. Not at this time. Votes are starting up in a minute. Will this panel be finished, or will they be held over until we are done?

The Chairman. They will be finished if there are no further questions at this point. You may proceed, or we can submit in writing at a later time.

Mr. Flake. I will conclude at this time.

The Chairman. Once again, we want to thank the panel for being with us today.

Mr. Flake. Thank you.

The Chairman. We will proceed with our third panel with the caveat that there are roll call votes expected on the Floor momentarily, and we may have to recess and come back, but I will call the panel to the table at this time.

Mr. Dennis Roller, Audit Manager, North Dakota State Auditors Office, Royalty Audit Section; Mr. A. David Lester, Executive Director, Council of Energy Resource Tribes; Mr. Michael Geesey, Director, Wyoming Department of Audit; and Professor Pamela Bucy, Frank M. Bainbridge Professor of Law, University of Alabama School of Law.

As the votes are now starting, I would suggest that the Committee take a half hour recess, and we will come back and start at the top of the list I just announced. The Committee stands in recess.

[Recess.]

The Chairman. The Committee will resume its sitting, and the panel will proceed with Mr. Roller going first.

We do have your prepared testimony. It will be made part of the record as if actually read, and you are encouraged to summarize and stay within the five minute limit.

STATEMENT OF DENNIS ROLLER, AUDIT MANAGER, NORTH DAKOTA STATE AUDITOR'S OFFICE, ROYALTY AUDIT SECTION

Mr. Roller. Mr. Chairman and Members of the Committee, I want to thank you for the opportunity to comment and share my views on some of the challenges faced by the Minerals Management Service and the North Dakota delegation.

Let me begin with a quick history of the North Dakota delegation. The North Dakota delegation was created in 1982 under the authority of Section 205 of the Federal Oil and Gas Royalty Management Act of 1982. For the past 25 years, the North Dakota delegation has performed compliance work on Federal mineral royalties paid in North Dakota with some very successful results.

As shown on Exhibit 1 of my written testimony, the North Dakota delegation from 1982 through 2001 collected over $26.6 million. During that same period, the costs of the North Dakota delegation were less than $4.2 million. That is almost $6 of revenue for every $1 spent.

Given the North Dakota delegation's success in the past, I now would like to discuss some of the challenges the MMS and the North Dakota delegation are facing. The first area is a state of misreporting for the MMS 2014s or the payment reporting document and the oil and gas operations report, the OGOR, or the production reporting document.

With the MMS reengineered system that went on line November 1, 2001, the MMS stopped doing any automated comparison of these two documents. Without any automated check, company reporting accuracy has deteriorated. This OGOR 2014 comparison process was a recommendation of the fiscal accountability of the nation's Energy Resources Committee of January 1982, commonly referred to as the Linowes Commission, which was a driving force for the creation of MMS.

Recommendation No. 5 of the Internal Control section of the Linowes Commission report states that the Federal royalty managers incorporate production data into the royalty management system in order to cross check the data with sales and royalty data for all leases each payment period.

The MMS did this comparison, commonly known as the AFS/PAAS comparison, prior to the implementation of the new system, and per the 2001 MMS budget justification document the AFS/ PAAS comparison process collected $56.2 million in additional Fiscal Year 1998 royalties.

Because of the level of misreporting, the North Dakota delegation requested the authority to perform volume and royalty rate automated verifications using a North Dakota developed tool. We have been performing these reviews using our tools since October 1, 2006.

Using this tool which compares the 2014s and OGORs, the North Dakota delegation has discovered countless reporting issues, nonpayment issues, missing document issues, nonreporting issues, two companies that just quit paying their Federal royalty obligation in North Dakota and well over $100,000 in additional royalties, not including the two companies that just quit paying their Federal royalties, all at a cost of less than $10,000.

The North Dakota delegation has taken on this comparison process, a process the MMS used to perform, at a time when North Dakota's funding has went from six FTEs and 4.5 FTE. The reporting issue goes to the core of having an effective royalty management program. Without correct reporting, the MMS does not know what their universe of receivables is and consequently cannot compare that universe to what was actually received, a basically principle for any business.

A second major area is the collection and information management system or CIMS. CIMS is the most current MMS collection tracking system that was made available to the North Dakota delegation in January of 2006. The previous collection tracking system was shut down in late 2001. The North Dakota collection information in the old MMS system was complete and accurate, and the North Dakota collection information in CIMS is incomplete and inaccurate.

A third area is the MMS reengineered system was not capable of calculating and billing late payment and collection interest until May 2003. Today the MMS says they are finally caught up on the backlog of interest. However, the North Dakota delegation has not received a report of the interest that has been billed from November 2001 through September 2006.

Finally, over the last five years what used to feel like a partnership of equals between the MMS and the delegations has now developed into something else. The MMS is increasingly issuing directives, requirements and mandates to delegations on almost every aspect of our delegation of authority with no negotiation or consultation.

In closing, the North Dakota delegation has been very successful in the past at collecting additional royalties owed from Federal lands. The level of misreporting and nonreporting has drastically increased. The collection information, CIMS, is incomplete and inaccurate. Interest since November 2001 is still an unknown, and, finally, the MMS has moved away from a partnership with the delegations.

This concludes my formal testimony. Thank you for the opportunity to appear before the Committee today, and I will be happy to answer any questions you may have on my oral or written testimony.

[The prepared statement of Mr. Roller follows:]

Statement of Dennis Roller, Audit Manager, North Dakota State Auditor's Office

Mr. Chairman and members of the committee, I want to thank you for the opportunity for me to comment and share my views concerning the wide array of challenges faced by the Minerals Management Service and State and Tribal delegations.

Let me begin with a quick history of the North Dakota State Auditor's Office Royalty Audit Section (ND delegation). The ND delegation was created in 1982 under the authority of section 205 of the Federal Oil and Gas Royalty Management Act of 1982 (FOGRMA). For the past 25 years, the ND delegation has performed compliance work on Federal mineral royalties paid in North Dakota, with some very successful results. I was an auditor for the ND delegation for over ten of those years and have been the audit manager for the ND delegation for the last three years.

As shown at Exhibit 1, the ND delegation from 1982 through 2001 collected over $26.6 million in additional Federal royalties. During that same period, the costs of the ND delegation were less than $4.2 million as shown at page 2 of Exhibit 1.

That's almost $6 of revenue for every $1 spent. For all State's that had a 205 delegation for 1982 through 2001 the total additional royalty collections were over $296.5 million, while costs were under $58.5 million. Exhibit 1 does not include any 202 Tribal delegation collections or costs as several Tribes prefer not to share that information. However, the ND delegation believes that 202 Tribal delegations have had similar success. Exhibit 1 is only through 2001 as that is the last date through which the MMS has accurate collection information. I will go into greater detail on the Collections and Information Management System (CIMS) later. These successful collection figures represent only the direct collections. For example, the ND delegation findings often have a residual financial effect due to future royalty payments being calculated correctly.

Given the ND delegation's success in the past, I now would like to discuss some of the challenges the MMS is facing that are recently limiting the efficiency of the ND delegation.

The first area is the state of misreporting for the MMS 2014s, the payment reporting document, and the Oil and Gas Operations Report (OGOR), the production reporting document. With the re-engineered system that was put in place on November 1, 2001, the MMS changed the property numbering system used by companies to report the 2014s. The MMS also stopped doing any automated comparison of these two documents. Without any automated check, company reporting accuracy has deteriorated. Our audits now often entail a reconciliation of every single payment made by a company for the review period in order to determine what the company intended to report and pay.

This OGOR-2014 comparison process also was a recommendation of the Fiscal Accountability of the Nation's Energy Resources Committee of January 1982, commonly referred to as the Linowes Commission, which was the driving force for the creation of the MMS. Recommendation #5 of the internal controls section (Chapter 3) of the Linowes Commission report states "That the Federal royalty managers incorporate production data into the royalty management system in order to cross check the data with sales and royalty data for all leases each payment period." (emphasis added) The MMS did this comparison prior to implementation of the re-engineered system and per the 2001 MMS budget justification document, this comparison process collected $56.2 million in additional FY98 royalties.

The deterioration of reporting has added a tremendous amount of hours to our audits. In order to combat this, the ND delegation requested the authority to perform volume and royalty rate automated verifications on October 1, 2005, as allowed for under the Federal Oil and Gas Royalty Simplification and Fairness Act of 1996 (FOGRSFA), see exerts from the request at Exhibit 2. The ND delegation was denied that request on January 20, 2006. However, the ND delegation was later granted the ability to perform limited scope compliance reviews using the comparison tool the ND delegation developed. The ND delegation tool uses the OGOR reported sales and the two known factors in the royalty equation, the Federal Governments allocation percentage and the royalty rate, to compare to what the company reported as owed on the 2014. The only remaining royalty equation factor is the unit value, which includes allowances and transportation. The ND delegation has been performing these limited scope oil volume and royalty rate compliance reviews since October 1, 2006. Using this comparison, the ND delegation has discovered countless reporting issues, non payment issues, missing documents issues, two companies that just quit paying their Federal royalty obligation in ND and well over $100,000 in additional royalties, not including the amount owed by the two companies that just quit paying, all at a cost of less than $10,000. The ND delegation has taken on this comparison process at a time when funding has been reduced from 6 FTE to 4.5 FTE and for FY08 the MMS has stated they will only fund the ND delegation at 4 FTE.

What the ND delegation has been finding with our comparison process is that companies are willing to correct their reporting when it is brought to their attention. MMS is no longer bringing it to the companies' attention and in fact we have had a company tell us that it can't be reported wrong because the MMS hasn't notified them that it's wrong.

This issue goes to the core of having an effective royalty management program. Without correct reporting, the MMS does not know what their universe of receivables is and consequently cannot compare that universe to what was actually received. This is a basic principle for any business.

A second major area is CIMS as I mentioned earlier. CIMS is the current MMS collection tracking system that was brought online January, 2006. The previous collection tracking system, CTS, was shut down in late 2001. The information in CTS agreed with the ND delegation's collection data. Four years later when CIMS comes online the data is inaccurate and incomplete. The MMS, on more than one occasion, has asked the delegations for help in correcting and reconciling CIMS. The ND delegation decided to use some of our limited resources to reconcile the CIMS information to our state data. After performing this reconciliation in mid 2006, it was determined that the reports that the ND delegation can generate from CIMS are not accurate. The reports do not reflect all the collections in CIMS for the ND delegation and there is no way for the ND delegation to generate a correct report.

Thirdly, the MMS re-engineered system did not have an interest module to bill late payment and additional royalty collection interest until May 2003. Today MMS says they are finally caught up on the back log of interest. However, the ND delegation has not received a correct report of the interest that has been billed from November 2001 through September 2006. The ND delegation did receive a report for interest from November 2001 through April 2006, but the report information was incorrect due to an error in the query.

On January 9, 2007, the MMS provided a report of interest for the quarter of October 1, 2006 through December 31, 2006. The ND delegation randomly reviewed two late payment interest billings from that report. For the first case, no late payment interest was actually owed. MMS billed late payment interest because the company misreported the sales month. The company reported the May 2002 sales month royalties as February 2002. So if the royalties were reported correctly, there would not have been interest owed.

The second case the ND delegation looked at was one for which the company claims they paid the royalty amount and MMS claims they didn't. This dispute centers on the matching or bookkeeping process the MMS has, a fourth area of concern. If a company reports on a 2014, that they owe \$100,000, but pays only \$90,000 the MMS matches the money as best they can. Normally the MMS would apply 90% to each of the detail lines of the \$100,000 report. In this case, the MMS determined that one sales month for one property for one product was not paid and thus billed the company interest for that one property, product, sales month. Putting aside the dispute over whether the amount initially was paid or not, the interest bill incorrectly calculated interest starting February 1, 2000 when the royalty was not due until March 1, 2000. Interest remains a concern for the ND delegation as the ND delegation has no report of interest billings from November 2001 through September 2006 and is concerned with the accuracy of the billings for October 2006 through December 2006.

Once again, the Linowes Commission provided a useful guideline for the MMS matching problem. In the summary section of the report, the commission stated "The Federal government should perform an oversight role. It must not waste its limited resources on tasks that are the industry's responsibility. In managing royalty collection, it should not remain mired in bookkeeping details that rightly belong to the lessee.'' The commission went on to state ''The oil and gas industry should carry out its obligation, as lessee, to pay royalties in full and on time. The industry, not the government, has primary responsibility for the detailed record keeping needed to assure that all royalties are paid.''

To further demonstrate the extent of how MMS has moved away from putting the accounting responsibility on the companies, the ND delegation has not issued an order to perform restructured accounting in over 5 years. In the past, whenever the ND delegation encountered a systemic problem, a problem for every test month or every property covered by a contract, the ND delegation requested the company to pay the additional royalties for the test months and then to perform restructured accounting (recalculate) the royalties for all the non test months. The MMS no longer is willing to sign orders to perform restructured accounting so instead the ND delegation has to test all months or project the non tested months. Unfortunately, if a projection is used, the dispute becomes the projection method rather than if additional royalties are actually owed.

The ND delegation concerns surrounding the financial system don't end with the matching or bookkeeping issue. Within the last year, the ND delegation has identified where an audit collection of $5,665 was distributed to ND at 50% of $5,640. An immaterial difference, but no explanation has been provided by the MMS for the difference. Another issue identified in the last six months is the MMS effectively borrowed ND's 50% share of the royalties for a property for up to three years and no explanation has been given. This issue is that several payments for ND's share of 2001 royalties for a property were backed out in October 2003, even though the company did not change their royalty reporting. Some of the amounts backed out were paid back three months later, some 12 months later and the final amounts were paid back 36 months later. Was the ND delegation paid late disbursement interest and how often has this occurred? Additional questions the ND delegation has not been provided answers for. Ideally the MMS financial system should be an automated process, but as MMS has recently stated to the delegations, there are way too many manual processes.

A fifth area that the ND delegation has found to be ineffective is MMS' Government Performance Result Act (GPRA) goals. The MMS goals are tied solely to reviewing a certain percentage of the revenue voluntarily paid by the companies. As the recent IG report points out, this goal results in only large companies being reviewed while there are hundreds of smaller companies that are never looked at. In fact, the ND delegation has found that it encourages the delegation to not look at companies that are severely underpaid. For example, the ND delegation put a company on our work plan that paid $0 in Federal royalties during the review period. The ND delegation got 0% credit towards the goal. However, the ND delegation knew this company owed over $100,000 of Federal royalties and had just failed to pay it.

Finally, the ND delegation has concerns about the working relationship between the MMS and the State and Tribal delegations. The delegation that has supported the MMS the most recently, has stated during STRAC (State and Tribal Royalty Audit Committee) only meetings that the limiting of the STRAC meetings to one a year by the MMS is in retaliation to STRAC going to congress with letters about MMS. At the last such STRAC only meeting, this delegation stated that it is a fact that the MMS is looking at legal ways of getting rid of STRAC. As you may be aware the MMS has reduced the number of STRAC meetings from quarterly to one a year and has eliminated the STRAC only portion of those meetings. See

Exhibits 3 through Exhibit 13 for individual delegation letters written to MMS in support of keeping the STRAC meetings. This is a great concern to the ND delegation as the State and Tribal delegations have had great success in collecting additional Federal royalties and protecting the United States Citizens mineral interests.

In closing, the delegations have been very successful in the past at collecting additional royalties owed from Federal lands. The ND delegation concurs with the recent IG report finding that the MMS management lacks reliable information to manage the compliance program. The collection information, CIMS is incomplete and inaccurate. The level of misreporting and non-reporting has drastically increased. Interest since November 2001 is still an unknown. The MMS is mired down in detailed bookkeeping which should be the responsibility of the industry. Goals encourage looking at high dollars and away from where it is likely that there is a significant percentage of additional royalties owed by a company. Finally, over the last five years the ND delegation has noticed a disturbing trend of the MMS in moving away from a partnership to a dictatorship in dealing with the delegations.

Thank you for the opportunity to provide written testimony to the Committee today. I will be happy to answer any questions you may have or provide further details and explanations surrounding any of these issues. My contact information is:

NOTE: The exhibits attached to Mr. Roller's statement have been retained in the Committee's official files.

The Chairman. Mr. Lester?

STATEMENT OF A. DAVID LESTER, EXECUTIVE DIRECTOR, COUNCIL OF ENERGY RESOURCE TRIBES

Mr. Lester. Thank you, Mr. Chairman. The Council of Energy Resource Tribes is honored to be invited to present its views to the Committee on this subject. I am David Lester, Executive Director for the Council of Energy Resource Tribes.

I have with me, who has helped the tribes since the enactment of FOGRMA in 1982, Mr. David Harrison and Virginia Boylan, attorneys who work with Indian tribes on these matters.

C.E.R.T., the Council on Energy Resource Tribes, was formed in 1975, and one of its first objectives was to get reform of how Indian royalty payments to tribes was handled by the Department of Interior. As a result of scandals, as Mr. Roller indicated, in 1982, Congress passed and the President signed into law the Federal Oil and Gas Royalty Management Act to reform, to correct the problems that were found in the late 1970s of the system as it existed then.

There are over 8,000 Indian leases, but there is no official way of finding out how many wells are drilled because the reporting is not at the well level. It is at the lease level. A lease can cover as little as 160 acres or as much as 250,000 acres, as little as a small handful of wells or thousands of wells. Compounding the data collection problem is that the numbers for Indian wells do not correspond to the numbering system used by the states and so we can't go and cross check well data from state records to Interior records.

I am going to try to highlight some of the major problems that the tribes have asked me to do, but also to request, if possible, that the Committee consider coming out to the field, perhaps allowing more detailed information to be gathered from tribes and individual Indian allottees. We would like to work with the Committee in gathering that information for you.

Compounding the urgency for the tribes in this matter is that Indian leases can be held past their primary term if the company or the operator continues to produce in the major portion paying quantities. In the sense the data is gathered at the lease level rather than at the well level, we have no way of performing that analysis to determine whether those leases that are being held in perpetuity are actually in compliance with the terms of the lease.

In short, even if everyone did their job, the system is not designed to fulfill its trust responsibilities to Indian individuals or to Indian tribes, a responsibility that is distinct in nature from its responsibility to public lands.

The FOGRMA was passed in 1982, but the final rules were not issued until 1990. That is because there was a tremendous attempt we believe, from our participation in the rulemaking process, of the agency to water down FOGRMA and to in effect reestablish the system that previously existed.

In short, the problems that FOGRMA sought to cure are with us today in the agency. Among those are the posted price problem. In many fields, the industry establishes a posted price which it uses to calculate royalties. Ten years ago the MMS abandoned the posted price on Federal leases, but continues to allow companies and operators to use posted price in calculating royalties owed to Indians and Indian tribes.

As far as we can see, in abandoning the posted price for Federal leases they have collected over $480 million more than they would have under the old system. We don't understand why they don't make the rule applicable to Indian tribes as well.

The internal presumption within the agency is that the operators and the royalty interest owners have common, in fact coterminous, interests. In fact, we have a number of cases of well defined adverse interests.

Major portion analysis. The leases require a major portion analysis; that is, that the operators are to pay royalties based on the highest price paid in the field on the majority or the major portion of that field. Unfortunately, MMS does not collect the data to do the analysis and so it is impossible for us to know whether the companies are complying with those terms of the lease.

In addition, as I said, we are getting lease level reporting rather than well reports and so we don't know what the paying quantities are on a well-by-well basis. Companies also take their standard deductions, which often result in the inability of the tribes to collect payment on liquids taken from natural gas, liquids that have market value.

Audit compliance versus compliance review. Indian audits that we have seen collect as much as 30 percent more than what has been paid. We have no confidence at all in compliance review and collecting what is owed to Indian tribes and Indian individuals.

Advice from the State Tribal Royalty Advisory Committee is often ignored. The agency seems to be in denial that there is any real problem that exists in the agency. They continue to use the honor system. When there is an overpayment claimed by a company it is automatically deducted from the payments.

When a tribe discovers an underpayment, it takes a procedure, an onerous procedure, for the tribe to collect the underpayment. When we do, the government requires us to sign a form that releases any claim that the tribe may have against the government, even when it is only about the company's payments, which leads us to believe that the government intends on reducing our ability to have our grievances rectified in Federal courts.

We want to say that this is not a partisan issue. This is not Democrats or liberals versus conservatives. This is about living up to the honor—the country and the tribes created the trust—about living up to the contractual agreements the parties have signed.

There are some things that are being done right. The lock box system seems to be working. The cooperative agreements, that is within MMS and BLM, bring tribal participation in. That is working and should be expanded rather than reduced, which is now the pressure to reduce the amount of money spent to cover the field inspections and audits.

That Farmington office that was earlier referred to, it was an example of what can be done when you bring all the agencies together, so the coordination and turf problems can be reduced. Unfortunately, the progress that was made is now dissipating because it no longer is held feet to the fire by the courts.

The decision making around separating all these Indian duties was made in 1982 by James Watt, which the tribes at that time opposed, but I want to make sure that we understand. It is not about where the boxes are. It is about getting the mission right and correcting the culture in the agency that seems to put us as adversaries; that the trustee and the beneficiary of the trust are inherently adversarial when in fact that should not be the case. The trustee should be our advocate and pushing for full payment, full reporting, full disclosure, complete transparency.

Free markets depend on transparency, and we are not given that opportunity for transparent transactions with respect to money owed us and production from our lands. I want to make sure that we know that.

I will finish with this statement. It is about family income. It is about capital for economic diversification and building a local economy. It is about funding essential governmental services at the tribal level. When we lose the revenues on nonrecoverable, nonrenewable resources we have cheated future generations.

It is not just about money. It is about what America stands for, and that is serving the people, and gaining the trust and confidence of the people. That is our government.

Thank you, Mr. Chairman.

[The prepared statement of Mr. Lester follows:]

Statement of A. David Lester, Executive Director, Council of Energy Resource Tribes, Denver, Colorado

Good Morning Chairman Rahall, Ranking Member Young and distinguished members of the House Committee on Natural Resources. Thank you, Mr. Chairman and Congressman Young for convening this hearing and the strong support you have shown for Indian Tribes and Native people over the years. I am David Lester, the Executive Director of the Council of Energy Resource Tribes (CERT) and an enrolled member of the Creek Nation of Oklahoma.

CERT was formed in 1975 in response to the energy crisis then gripping the United States and our national need to increase domestic production of oil, gas and hard rock mineral resources. CERT is an organization formed by Indian Tribes to work for Indian Tribes and is a true Inter-Tribal organization. CERT's membership includes 53 Federally recognized Tribes and three First Nations from Canada.

Thirty-two years after its formation, CERT's mission remains the same: to support its member Tribes in their efforts to develop management capabilities and use their energy resources to help build stable, diversified, self-governing economies according to each Tribe's own values and priorities. CERT's programs include policy advocacy, technical assistance, education and capacity-building partnerships. As part of capacity-building and information sharing, CERT disseminates knowledge through www.certredearth.com and other media.

I very much appreciate the opportunity to present testimony to the Committee on a matter of great concern to our member Tribes. Because CERT is a tribal organization, my testimony will focus on mineral activity on tribal trust lands and transactions involving leases between Indian Tribes and their private sector partners. Individual Indians whose lands are leased for mineral activity, and to whom the Minerals Management Service (MMS) also owes a trust responsibility, have a related but distinct set of issues that I trust the Committee will explore through the testimony of other witnesses. Similarly, my testimony is related primarily to oil and gas leases, rather than issues related to royalties from coal production even though MMS also collects royalties from tribal coal leases. We do understand there are myriad problems with undervaluation and underreporting on hard rock mineral leases that are similar to oil and gas development.

I am pleased to be accompanied today by David Harrison, an attorney from Albuquerque, New Mexico. Mr. Harrison is a former director of the Bureau of Indian Affairs trust office and has long experience in dealing with MMS and the issues facing Tribes in the area of royalty accounting. Mr. Harrison is here to assist me in answering questions of a technical nature that the members may have.

Role of the Federal Mineral Management Service

At the outset, it is important to note that because the United States (U.S.) holds legal title to the vast mineral estate that lies below Indian tribal lands, the U.S. is bound to act as trustee for the benefit of the Tribes and tribal members.

Federal laws such as the Indian Mineral Leasing Act and the Indian Mineral Development Act require Federal approval of the terms and conditions under which Indian energy resources are developed. The Federal government invariably has some role in monitoring such development and in overseeing the payment of royalties to the Indian resource owner.

In this situation, Indian Tribes are often dependent on the MMS in the Department of the Interior to provide accurate and timely accounting and collection of royalties from gas, coal, and oil companies engaged in mineral activity on tribal trust lands. However, the Federal Oil and Gas Royalty Management Act also provides a mechanism that allows Tribes to enter into contracts with the MMS to perform audits of their own leases under MMS supervision. A number of energy producing Tribes have entered into cooperative audit agreements with the MMS and have been conducting royalty audits for many years.

Just as participation in audit review reflects greater tribal involvement in resource management, other recent developments further illustrate the evolving nature of Federal Indian law and policy regarding tribal energy resource development. The Energy Policy Act of 2005 includes as Title V the Indian Tribal Energy Development and Self-Determination Act, which authorized Indian Tribes to negotiate and enter into with the Secretary of the Interior "Tribal Energy Resource Agreements" (TERAs) for purposes of energy development, land management, and environmental regulation on tribal lands. The major element of such a TERA, once approved by the Secretary, is that the Indian Tribes themselves will have the decision-making authority over a host of energy and related matters that are currently relegated to the U.S.

Indian tribal regulatory capacity is ever-increasing and that, coupled with the Tribes' business sophistication, will bring a new day in the near future when the advances in tribal authority made possible by Title V might be expanded to include lease monitoring and royalty verification functions that are currently relegated in large measure to the MMS.

The State and Tribal Royalty Audit Committee Experience

CERT believes this hearing is not just timely, it is, in truth, long overdue. Some of the Tribes from whom CERT solicited input for today's hearing have reported long and sometimes very difficult tales about their dealings with the MMS. Although issues of concern to Tribes are well known to the officials at MMS, the resolution of these concerns has never been a priority of that agency and these concerns continue to reside in the never-never land of the Federal bureaucracy.

The experience of the State and Tribal Royalty Audit Committee (STRAC) is illustrative. STRAC was established in 1986 as an association of royalty auditors for states and Tribes, who had entered into delegation agreements or cooperative agreements with the MMS to perform royalty audits. Members of STRAC have long been committed to excellence in auditing, and the STRAC organization has provided a forum for discussion of common audit issues, peer review, and development of policy recommendations for improvements in valuation and auditing policies and practices. In recent years, many members of STRAC have expressed concern about emerging MMS policies and practices that may adversely affect the collection of royalties by the Federal government, states, and Tribes under Federal and tribal leases. Some Members of Congress have also expressed concern about these matters.

Following up on a February 2006 letter from STRAC to the MMS, in April 2006, seven Members of Congress, Representatives Carolyn Mahoney, Henry Waxman, George Miller, Raul Grijalva, Edward Markey, Maurice Hinchey, and Rahm Emanuel, wrote to MMS Director R.M. ''Johnnie'' Burton requesting information about MMS's compliance review program. In particular, the letter sought information and analysis from MMS about the cost and effectiveness of ''compliance reviews'' (CRs) compared to more traditional audits covering leases of Federal and Indian tribal lands. See attached April 3, 2006 letter. Compliance reviews are a short-cut method of reviewing unverified company reports to determine company compliance with royalty payment obligations.

In May 2006, Director Burton responded and indicated that, among other things, in the five-year period between 2000 and 2005, CRs grew from 34% of agency costs to 63% of agency costs as compared to more traditional audits. Director Burton stated that ''[w]ith over 27,000 producing Federal and Indian mineral leases under our jurisdiction, there are simply too many properties to rely on the traditional audit approach alone.'' See attached May 17, 2006 letter. Director Burton defended the agency shift to CRs by comparing historical audit collections with the collections obtained under CRs.

The Southern Ute Indian Tribe, a member of the STRAC, received a copy of Director Burton's May 2006 response and, in turn, sent a letter to Director Burton taking exception to the inaccurate information and representations contained in Burton's response. The Tribe was very concerned that, in defending the shift away from audits to CRs, Director Burton had significantly understated the amounts actually collected by states and Tribes through audits. The Tribe requested that Director Burton and MMS dedicate additional staff and initiate corrective action and communications with the congressional recipients of the erroneous information. See attached June 21, 2006, letter.

CERT was encouraged to learn that in July 2006 the Southern Ute Indian Tribe was notified that additional MMS staff would be dedicated to rectify the erroneous collections information. We hope that Congress, as well as the states and Tribes that received the May 2006 letter from Director Burton will soon receive correspondence informing them of the agency's corrective action, along with updated information. Because this information appears to be a key justification for the shift from traditional audits to CRs, its accuracy is important in any meaningful discussion of the relative merits of CRs and audits.

CERT trusts that this Committee will provide the guidance that is needed to the MMS to inspire Tribes to ensure that the degree of monitoring, audits and accountings and royalty payments owed them are, without question, accurate and will be paid in a timely manner.

In addition to the need for accurate accountings and disbursements, Indian Tribes need assurance that when disputes inevitably arise between Tribes and their private sector partners, the Department of the Interior will act to fulfill its obligations to the Tribes and will not be an active impediment to settlements between Tribes and the companies. Tribal officials have asked that CERT bring to the Committee's attention the following issues of concern. Many of these matters are interrelated but I have tried to organize them so that the potential solutions to each are clear.

The Honor System for Oil Valuation

At the outset, I want to impress upon the Committee that the current legal framework and business model under which companies report their own production, and therefore the royalties owed tribal resource owners, can only be described as "the honor system", and it remains a major problem for Tribes. The reality is that the conclusions of the Linowes Commission in 1982 that companies were on an "honor system" is still true today, at least with regard to royalties payable on tribal leases. You will recall that the Linowes Commission was established to investigate allegations of Federal mismanagement of oil and gas royalties. Its findings led the Congress to enact the Federal Oil and Gas Royalty Management Act (FOGRMA).

FOGRMA, in turn, led the Department of the Interior to engage in what we in Indian Country have grown accustomed to witnessing over the years in the face of malfeasance or nonfeasance from our Federal trustee: an internal "reorganization" that resulted in no substantial improvement in MMS performance. Consequently, hundreds of millions of dollars were spent on computer systems and automated reporting regimes, scores of pages of fine print regulations were promulgated to deal with valuation and accounting issues. Sadly, if the Linowes Commission were reconstituted, it would report that the minerals industry is still operating on the honor system, at least with respect to tribal mineral leases.

Since 1998, the Navajo Nation and other Indian Tribes have been urging the MMS to issue regulations governing oil royalty valuation for Indian tribal mineral leases. MMS has issued regulations on oil royalty valuation for Federal leases, but the regulation governing Indian tribal mineral leases has been withdrawn. The result is that oil production on tribal leases is still valued under rules that went into effect on March 1, 1988. As the Committee might guess, the oil marketing practices of the industry were vastly different 19 years ago than they are today but Indian Tribes and Indian allottees still have to rely on "posted oil prices" that are set by the industry itself. These prices do not represent fair market value, and the Committee should understand this as it moves ahead with its oversight efforts.

The honor system unfailingly results in underpayment of amounts owed, and sometimes the underpayments are huge. In the 1980s, MMS proposed as a conclusive presumption that an operator in an arms-length transaction with an Indian Tribe was dealing in the best interest of the royalty owner as well as himself. The royalty owners, including Tribes, were successful in persuading the MMS to reject the idea of a conclusive presumption. The MMS abandoned the notion that the interests of producers and royalty owners are co-terminus, and it also abandoned the notion that posted prices represent the true value of crude oil. The MMS changed the regulation with respect to production of oil from Federal leases, but has not acted with respect to Indian tribal royalty owners. Thus, the honor system persists, but only to the detriment of Indian royalty owners. We believe the Committee will agree that this is unconscionable and needs to be revisited.

Release of Claims Against the United States

Because the U.S. holds title to Indian-owned energy resources such as oil, gas, and coal, when either the U.S. or the Tribe discovers that a company has underpaid significant amounts of royalties, the MMS makes demands for compensation against the company. When settlements are reached, MMS lawyers routinely insist that settlements contain hold harmless provisions protecting the U.S. from any other claims relating to that same production that are not covered by the particular settlement agreement.

In all cases, it is the task of the MMS to value and collect royalties accurately. Only when it fails to do so and the underpayment is discovered, usually by the Tribe, does this issue arise. While the U.S. retains responsibility for the enforcing the terms of the leases, it does not accept any liability for failure to do its job. Claims may be unknown to the Tribe at the time of settlement and it is an onerous burden to require Tribes to pursue and prosecute these issues. CERT recommends that the Committee direct the MMS to accept settlement agreements without also requiring release by Tribes of collateral claims against the U.S.

Major Portion Pricing

Standard Bureau of Indian Affairs Indian tribal mineral leases require that royalties be paid on not less than the highest price paid for a "major portion" of like quality production on the same field or area during the production period. In trying to apply this requirement over the years, it has been agreed to apply it only to armslength agreements. The problem is that the MMS has never collected the information necessary to perform any analysis of either "major portion" or what constitutes an "arms-length" agreement. Whether this failure to act is caused by understaffing or by simple neglect, the fact is that Indian tribal leases continue to be significantly undervalued and thus underpaid.

The data collected by the MMS regarding production under Indian tribal leases is not sufficient to support or fulfill the trust obligation of the U.S. Revenue collected by the MMS from Indian leases is only about two percent (2%) of the total production income from all leases on Federal or tribal lands. It is our view, however, that this is the most important percentage of collections because it represents only a fraction of the true value of the tribal resources that, once extracted, are lost forever. CERT respectfully suggests that this Committee take the necessary steps, whether in the form of additional funding for the MMS or in clarifying the law, to ensure that Indian Tribes (and individual Indians) are paid accurately and fully for the value of their resources. Once the resources are extracted, they cannot be replaced.

Negative Payments

Federal offshore mineral leasing laws prohibit companies from claiming recoupment of overpayments for a prescribed number of years. This is not the case with leases on Indian lands. Thus, if a company decides in 2007 that it overpaid for minerals extracted in 2006, it can simply reduce currently owed payments by the amount it claims it overpaid the year before. These unilateral adjustments are often made without the Tribes' knowledge and, more important, without review or oversight by the MMS. Similarly, a recoupment might occur at any time because a mineral lessee might seek to recoup overpayments against 2007 production from years ago. If an Indian Tribe is not in a position to know about these calculations and private company decisions, negative payments just happen. The current system, if left untouched, will continue to fail tribal resource owners.

Cooperative Agreements And Lockboxes

On a positive note, the MMS has entered into cooperative agreements with several Indian Tribes under which companies simultaneously issue production reports to the U.S. and to the Indian tribal mineral owner. In these situations, royalty payments are sent directly to the Tribe's financial institution. At one time, it took some Tribes years to get duplicates of lease payment reports, and, consequently money transfers and confirmations were not made in a timely fashion, which resulted in confusing and often wildly inaccurate bookkeeping on the underlying production.

Not all Tribes have cooperative agreements, but those that do seem genuinely pleased with the timeliness of the payments and the lockbox system. CERT recommends that the Committee encourage these types of cooperative agreements and expand their use.

Resistance to well-by-well Reporting

A mineral operator can keep a lease in force in perpetuity, even after the primary lease term has expired, as long as the lease is producing ''in paying quantities'' on a major part of the lease. This is true whether the lease is for 160 acres or 250,000 acres, and regardless of the number of wells. If the operator does not report on a well-by-well basis, however, there is no accurate way to know whether the production is ''in paying quantities.'' When the ''major part'' of the lease is not producing ''in paying quantities,'' the wells must be shut in and the lease expires. Once wells are shut in, the only way to resume production is to negotiate a new lease. To ensure correct pricing, the mineral operator's production on a well-by-well basis should be recorded.

Audit Compliance

The MMS has established an audit compliance and review program related to its royalty accounting system. At the outset, most Tribes believe that the MMS's reliance on voluntary submission by the company of their respective oil and gas sales contracts is misplaced. CERT Tribes agree with such an assessment, based on the past and present performance of the industry. Further, technical compliance tools that are being used successfully by Onshore and Indian Compliance Asset Management organizations at the Minerals Revenue Management offices simply do not work from the desk stations of auditors in remote tribal headquarters. In these cases, Tribes rightly place little faith in data generated by MMS' system to perform audits.

Conclusion

At least one person who audits company-reported oil and gas payments for a Tribe in the Southwest routinely reports a 30% underpayment for natural gas produced on tribal lands. Any system that allows such underpayments, especially underpayments of this magnitude, is immoral and untenable. As with so many other activities and functions in other areas that are supposed to be performed by the U.S., the only way some Indian Tribes have been able to monitor the payments is by doing it themselves.

Some Tribes actually monitor payments under contracts with the MMS. However, monitoring is an MMS responsibility. If Tribes are routinely finding these kinds of underpayments, it makes one seriously question the ability or willingness of the MMS to account for the other 98% of mineral leases where royalty payments are supposed to go to the U.S. Treasury.

On behalf of the member Tribes of CERT, I again thank you for the opportunity to appear before you today on this very important matter and am happy to answer any questions you might have.

NOTE: Attachments to Mr. Lester's statement have been retained in the Committee's official files.

The Chairman. Mr. Geesey?

STATEMENT OF MICHAEL GEESEY, DIRECTOR, WYOMING DEPARTMENT OF AUDIT

Mr. Geesey. Mr. Chairman, Members of the Committee, I would like to thank you for the opportunity to talk about the Federal royalty management program. I am Mike Geesey with the State of Wyoming, Department of Audit.

I am here today at the Chairman's invitation. As the Chairman is well aware, Wyoming has a large amount of onshore mineral production where Federal royalties are paid. Wyoming's Department of Audit contracts with the Federal government to audit royalty payments from that production.

I offer the following comments from a state's point of view. In 1998, KPMG Peat Marwick presented an independent report titled Cost Allocation of Royalty Management

Program, A Function Performed by the Minerals Management Service. The role of KPMG was to present, among other analysis, an independent report on the allocation of existing Federal and state audit resources.

The conclusion in short was that the state audit programs are resource effective and cost efficient. However, some states were underfunded and other states were overfunded, thus necessitating the need for a reallocation of the resources.

Wyoming has worked with the Minerals Management Service to allocate those limited resources to match the risk and oversight needed to ensure a proper onshore Federal royalty payment compliance. In 2000, we received about 20 percent of the funds budgeted for all state audit programs with over 40 percent of the onshore royalty payments being made from Federal leases in Wyoming.

Last year, 2006, Wyoming received 30 percent of the state audit funding with over 50 percent of the onshore payments being made from leases in Wyoming. I believe there still needs to make additional reallocation of the present budgeted funds to balance the funding of royalties at risk. Given Wyoming's current Federal production, the reduction in any funding could potentially place some of the Federal royalty payments at risk.

The Department of Audit is proud of the role that we have played in helping the Minerals Management Service with its onshore royalty management program. We have consistently strived to improve our audit program. Likewise, the Minerals Management Service's efforts to do the same includes the implementation of a compliance review program.

The State of Wyoming Department of Audit has always had a professional working partnership with the Minerals Management Service and sees the compliance review program as an additional tool in the audit process. Wyoming acknowledges that this program complements the current onshore royalty audit program by enhancing the risk assessment process and providing early detection of noncompliant Federal royalty payments.

This process has not and will not reduce the number of full audits we plan to undertake. It only helps focus our limited resources toward Federal royalty payments at risk.

Wyoming is disappointed to read that the Minerals Management Service's 2008 budget is considering amending Section 35 of the Mineral Leasing Act to syphon off administrative costs for the singular act of calculating 50 percent of the total royalties paid in a state and writing a check for that statutory payment. Those costs, using 2006 onshore royalties, would amount to over $44 million in reduced Federal royalty payments to the states.

Presently this is in conflict with Section 503 of the Mineral Payment Clarification Act of 2000. Wyoming sees no reason to return to the controversial net receipt policy.

I thank you for giving me the opportunity to make some comments.

Thank you.

[The prepared statement of Mr. Geesey follows:]

Statement of Mike Geesey, Director, Wyoming Department of Audit, Cheyenne, Wyoming

Mr. Chairman and members of this committee, I would like to thank you for the opportunity to talk about the federal royalty management program. I am Mike Geesey, with the State of Wyoming, Department of Audit. I am here today at the Chairman's invitation. As the Chairman is well aware, Wyoming has a large amount of onshore mineral production where federal royalties are paid. Wyoming's Department of Audit contracts with the Federal Government to audit royalty payments from that production. I offer the following comments from a State's point of view.

State Funding of Federal Royalty Audits

On January 23, 1998 KPMG Peat Marwick LLP (KPMG) presented an independent report ''Cost Allocation for Royalty Management Program (RMP) Functions Performed by Minerals Management Service (MMS).'' The role of KPMG was to present, among other analysis, an independent report on the appropriate allocation of existing federal and state audit resources. The conclusion, in short, was that state audit programs are resource effective and cost efficient. However some states were over funded and other states were under funded, thus necessitating the need for a reallocation of audit resources. Wyoming worked with the MMS to allocate those limited resources to match the risk and oversight needed to ensure proper onshore federal royalty payment compliance. In 2000, we received about 20% of the funds budgeted for all the state audit programs, with over 40% of the onshore royalty payments being made from federal leases in Wyoming. Last year 2006 Wyoming received 30% of the state audit funding with over 50% of the onshore payments being made from leases in Wyoming. I believe their still needs to be an additional reallocation of the present budgeted funds, to balance audit funding with royalties at risk. Given Wyoming's current federal production, any reduction in funding could potentially place some federal royalty payments at risk.

Compliance Review Program

The Wyoming Audit Department is proud of the role we have played in helping MMS with its onshore Royalty Management Program. We are constantly striving to improve our audit program. Likewise, Mineral Management Service's efforts to do the same, includes the implementation of a compliance review program. The State of Wyoming, Department of Audit has always had a professional working partnership with the MMS and sees the compliance review program as an additional tool in the audit process. Wyoming acknowledges that this program, compliments the current onshore royalty audit program, by enhancing the risk assessment process and providing early detection of noncompliant federal royalty payments. This process has not and will not reduce the number of full audits we plan to undertake. It only helps focus limited resources towards federal royalty payments at risk.

MMS Budget Concerns

Wyoming is disappointed to read MMS's 2008 budget is considering amending Section 35 of the Mineral Leasing Act (30 U.S.C. 191) to siphoned off administrative costs for the singular act of calculating 50% of total royalties paid in a state, and writing a check for that statutory payment. Those cost using 2006 onshore royalties would amount to over $44 million in reduced federal royalty payments to the states. Presently this would conflict with section 503 of the Mineral Payments Clarification Act of 2000. Wyoming cannot see any reason to return to a controversial policy of "net receipt sharing."

STATEMENT OF PROFESSOR PAMELA BUCY, FRANK M. BAINBRIDGE PROFESSOR OF LAW, UNIVERSITY OF ALABAMA SCHOOL OF LAW

Ms. Bucy. I would like to thank the Committee for the invitation to appear before the Committee. I am a law professor. I was a Federal prosecutor prosecuting almost exclusively white collar crime, and now I teach, publish and speak on the Civil False Claims Act.

In my statement I would like to address four points. First, the goal of the False Claims Act; second, how that goal is affected when government employees who get the information through their employment bring lawsuits under the statute;

Third, how the statute could be amended to remedy the problems that are created when such employees bring lawsuits; and, fourth, how to better remedy the problem that Mr. Maxwell and other government employees run into when the agency that they work for is not responsive to the problems. There is a better solution for doing that rather than going through the False Claims Act.

Whether you like the False Claims Act, which the Department of Justice does, or you hate it, which most industry does, there is no controversy at all that it is effective. It is recognized as the premiere fraud fighting tool of the Federal government. Last year, of course, Congress passed legislation requiring all of the states to pass their own Civil False Claims Act that mirrored the Federal statute.

The statute is unusual in that it allows a private person known as a qui tam relator to bring the lawsuit, as well as the Federal government. This private person need not be damaged or affected at all by the fraud. All they have to have is information that someone else has filed false claims with the government, and that allows them to bring the lawsuit.

That person, the relator, is guaranteed a percentage of the judgment, so, for example, even though Mr. Maxwell does not know what percentage he will get, he is guaranteed 15 percent under the statute. The False Claims Act after the 1986 amendments is now silent on the question that this committee has raised, and that is whether government employees can serve as qui tam relators.

The courts have weighed in on this, and there is a split in the Circuits. Two of the Circuits say government employees do qualify, and two of the Circuits say they do not. On each side there is an en banc opinion, so clearly this is ripe for the U.S. Supreme Court to take.

All of the courts that have addressed it, even the ones that say that the statute does not allow the court to preclude the government employees from serving as relators, all of the courts agree that allowing government employees who get the information through their employment frustrates the goal of the statutes, so that is the point that I would like to cover is how that does frustrate the goal.

I think it is also worth noting that every single court that has addressed it has asked that Congress remedy this problem because it is an ambiguity in the statute.

The major goal of the Civil False Claims Act is to bring information of fraud to the Federal government. Fraud is hard to investigate. You need an insider. You need somebody who can tell you who did it, where the documents are and how it happened. The False Claims Act incentivizes insiders to come forward.

But, to allow a government employee who gets that information through their government duties is parasitic, and that is not the goal of the False Claims Act, which is to bring information the government otherwise cannot get.

It also creates seven perverse incentives, and this is how the courts talk about it, even the ones that say they don't have the authority and Congress has to do it. The first incentive is that it is going to give a government employee who is going to be allowed to bring a lawsuit under the Civil False Claims Act, the first incentive is to conceal information, to conceal information about the fraud from their superiors, from their co-workers and from government prosecutors so that they can bring the lawsuit themselves.

The second incentive is to race to the courthouse to beat the government in filing the house. The third incentive is to prematurely disclose information which could cause the parties that are being investigated to destroy documents or to get their stories together.

The fourth incentive, contractors will be deterred from being honest with auditors because they have to wonder am I providing this confidential proprietary information to this person so they can go file a lawsuit on their own behalf or so that we can get this problem remedied to help the taxpayers.

Fifth incentive, the auditors who are often going to be in the position to become these qui tam relators are also the key government witnesses in any prosecution. Those witnesses' credibility is destroyed once they become a plaintiff in a case.

Sixth, the public confidence is destroyed whenever you have to worry about your government official taking care of themselves rather than their fiduciary duty to the public. Seventh, of course it is inconsistent with all of the conflict of interest requirements for government officials that they not use their public position for private gain.

How to fix this. It is easy to fix. Congress passed a qui tam provision to prevent bank fraud that is actually mirrored after the False Claims Act and in that did what needs to be done here. It is 15 words that disqualify a current or former government employee who discovered or gathered the information in the qui tam lawsuit in whole or part while acting within the course of their government employment. That is all it would take to fix this problem.

There is still the issue of what to do if you are working for an agency that has not responded, and I would suggest that that has already been taken care of in the context of public companies. The Sarbanes-Oxley statute gave the impetus to the SEC and to the ABA for lawyers to pass reporting up and reporting out requirements.

That would require that there be an established protocol to report within your agency, to report to the IG and report to a fraud expert in the Civil Division at Department of Justice. Those reportings would need to be legitimate reportings, not a canceled appointment with the IG as in Mr. Maxwell's case.

This concludes my opening remarks. I have all of this in my written statement, and I will be more than happy to answer any questions.

[The prepared statement of Ms. Bucy follows:]

Statement of Pamela H. Bucy,[10] Frank M. Bainbridge Professor of Law, University of Alabama School of Law

Part One of this Statement provides an overview of the civil False Claims Act. Part Two discusses court decisions and policy issues in allowing government employees to qualify as ''relators'' under the civil False Claims Act.[11] Part Two also addresses the recent case of United States ex rel. Bobby Maxwell v. Kerr McGee Oil & Gas Corporation.[12] Part Three proposes an amendment to the False Claims Act to clarify when government employees should qualify to serve as relators under the civil False Claims Act.

I. Overview of the Civil False Claims Act

The False Claims Act,[13] first passed in 1863,[14] and amended several times since,[15] most dramatically in 1986,[16] is recognized by the United States Department of Justice as its ''primary'' civil enforcement tool.[17] The Act grows out of a long tradition of using private parties to supplement law enforcement efforts.[18]

In American jurisprudence today there are a number of actions that private parties may bring alleging that a defendant has violated some federal or state law.[19] To the extent these actions supplement the efforts of law enforcement in detecting, proving and deterring lawbreaking, the private parties who bring them serve as ''private attorney generals.'' In almost all of these actions, the private party who brings the action has been personally injured by the defendant's conduct. The False Claims Act is unique among these actions because it allows a private party who has not been personally injured to bring the FCA action alleging violation of public laws by the defendant.

The FCA provides that a person who believes that he has information and evidence that someone else (individual or company) has filed false claims against the federal government, may file a lawsuit making such allegations (termed a ''qui tam action).[20] This plaintiff (termed a ''relator'') is required to file his lawsuit under seal (not even serving it on the defendant). The relator is also required to give a copy of the lawsuit to the United States Department of Justice, along with a written report of ''all material evidence and information'' the relator possesses.[21] The lawsuit stays under seal, often for two years or more, to allow DOJ to fully investigate the charges made by the relator. The secrecy provided by sealing the complaint not only protects a defendant's reputation if the relator's information amounts to nothing, but also facilitates DOJ's further investigation of the relator's information.

At the conclusion of its investigation, DOJ decides whether it will intervene in the lawsuit as an additional plaintiff. If it does, DOJ assumes ''primary responsibility'' for the case although the relator remains as a plaintiff and is guaranteed a participatory role.[22] In some cases, DOJ handles the entire case after intervening; in others, relators work hand-in-hand with government prosecutors. In some cases, relators and their attorneys assume the bulk of the investigative and litigative duties.[23]

If DOJ does not join the lawsuit, the relator may continue pursuing the case, litigating it alone.[24] Even if DOJ does not join a relator's case, it retains authority over the relator's lawsuit in several ways: DOJ monitors the case and may join it at any time, even for limited purposes, such as appeal;[25] DOJ may settle or dismiss a relator's suit over the relator's objections as long as the relator has been given an opportunity in court to be heard;[26] DOJ may seek limitations on the relator's involvement in the case,[27] or seek alternative remedies (such as administrative sanctions) in lieu of the relator's lawsuit.[28]

If the government joins the relator's case, the relator is guaranteed at least 15% of any judgment or settlement and the court can award more—up to 25%. If the government does not join the lawsuit, the relator is guaranteed 25% and could receive up to 30%.[29] The amount within the statutory award depends upon the relator's helpfulness to the government. Because the FCA's damages and penalty provisions tend to generate exceptionally large judgments, relators' percentages involve substantial sums[30]

The case of United States ex rel. Alderson v. Quorum Health Group,[31] demonstrates how the FCA works. It is typical in that it shows the steps of a FCA qui tam action. It is atypical because of the unusual contribution made by the relator to pursuing the case; in this respect, Alderson exemplifies the FCA working to its fullest potential.

In the 1980s, Alderson was the Chief Financial Officer at North Valley Hospital in Whitefish Montana. He had been so employed for six and one-half years.[32] In August, 1990, Quorum Health Group took over as the management company for the hospital. Soon thereafter a Quorum representative instructed Alderson to prepare two Medicare cost reports. Hospitals that participate in the Medicare program by treating Medicare patients must submit annual cost reports. These are lengthy, detailed reports that provide extensive information about a hospital's costs.[33] Alderson was told to prepare an ''aggressive'' cost report to submit to Medicare, and a ''reserve'' report to be used internally.[34]

Alderson refused to prepare the two inconsistent reports. He was terminated four days later.[35] Within months Alderson filed a wrongful termination suit.[36] During depositions regarding his termination, Alderson learned of additional irregularities in Quorum's cost-reporting practices. He sought documents that would shed further light on such practices and engaged a forensic accounting expert.[37] In 1992, two years after his termination by Quorum, Alderson filed a pro se FCA qui tam complaint alleging that Quorum's cost reporting practice defrauded the Medicare program. As required by the FCA, Alderson provided the federal government with a copy of his complaint and a written statement of the information and evidence he had gathered supporting the charges in his complaint.[38]

Unable to find another job after being fired from North Valley Hospital, Alderson and his family suffered financially for years after his termination. His family was forced to move from its comfortable home to a cramped apartment in another town. They used the college savings they had accumulated for their two teenage children.[39]

For nine years after he filed his pro se FCA complaint, Alderson spent thousands of hours working on his FCA case,[40] retained two different law firms to represent him in the action,[41] and either by himself or with his attorneys, met often with DOJ attorneys and/or investigators, mostly in Washington D.C., and at his own expense.[42] At these meetings, Alderson explained how Quorum's reserve cost report practice defrauded the Medicare Program.[43] When DOJ attorneys expressed concern about a legal theory to support an FCA case, or what they viewed as weak evidence or minimal damage,[44] Alderson addressed their concerns.[45] The forensic accountant Alderson retained, and continued to pay, met with DOJ officials in Washington D.C. to assist Alderson in explaining the fraud to DOJ attorneys.[46]

Working with the DOJ attorneys and investigators, Alderson identified voluminous documents that government investigators should subpoena from Quorum.[47] At DOJ's request, he reviewed the documents obtained by subpoena.[48] These were extensive: eight boxes of more than 11,000 records from 197 hospitals for seven years. For one year, working alone, Alderson analyzed the records and prepared a spread sheet summary of relevant cost reserve information. He culled a set of 2,500 documents that corroborated specific reserve information and presented his summary, spreadsheet and relevant documents to DOJ.[49]

Seven years after Alderson filed his action[50] DOJ agreed to intervene in Alderson's lawsuit, but only after receiving ''assurances from Alderson's counsel of their ability and willingness to commit the necessary resources to the case and to undertake the principal role in prosecuting the litigation.''[51]

The case ultimately settled for $85.7 million. Alderson's share was $20.6.[52] The average relator's award, when the government intervenes, is 16 % of judgment recovered.[53] When awarding Alderson an unusually large award of 24 %, the Court looked to the FCA, its legislative history, DOJ Guidelines for Relator's Award,[54] Alderson's persistence,[55] expertise, the personal sacrifices he made to help the government,[56] and the significant contribution of Alderson's counsel in pursuing the case.[57] Alderson's attorneys were awarded $2.7 million in attorneys fees pursuant to the FCA's requirement that culpable defendants should pay ''reasonable attorneys' fees and costs.''[58]

The court that determined Alderson's share described Alderson's contribution to the case: ''[t]he record graphically demonstrates Alderson's profound personal and professional commitment to success in this litigation. His commitment manifested itself in his persistent labors and those of his attorneys and accountants, all of whom contributed mightily both before and after the United States intervened.''[59]

II. Government Employees as Relators under the Civil False Claims Act

A. Policy Reasons and Court Decisions

Because the civil False Claims Act (FCA) is widely recognized as the most powerful tool available for pursuing fraudulent government contractors and because government employees often learn of such fraud in the course of their duties, the issue of whether government employees qualify to sue under the FCA is significant, arises often, and needs to be clarified by Congress.

Given the consensus by courts and policy makers that allowing government employees to bring FCA actions creates serious enforcement and fairness problems, and courts expressed frustration in their ability to address these problems vis à vis Congress's authority to do so, Congressional attention is needed. The FCA is an outstanding fraud-fighting tool but to remain effective it needs clarification on the issue of whether government employees qualify as FCA plaintiffs.

The courts that have addressed this issue disagree on the question whether the FCA, in its current form, permits government employees to bring suit under the FCA. The First[60] and Ninth Circuits have held that government employees who obtain information in the course of their official duties do not qualify to sue under the FCA.[61] The Tenth and Eleventh Circuits have held that government employees do qualify.[62] All of these courts, however, uniformly agree that it is poor public policy to allow government employees to bring lawsuits under the FCA. The only point on which they disagree is whether the courts, or Congress, should remedy the problem. As the Tenth Circuit noted in reluctantly holding that government employees were eligible to bring FCA lawsuits: ''Although there may be sound public policy reasons for limiting government employees' ability to file qui tam actions, that is Congress' prerogative, not ours.''[63]

The courts are right. There are significant policy reasons that government employees who obtain information about fraud in the exercise of their official duties, should not qualify to bring FCA lawsuits. Termed ''perverse incentives'' by the U.S. Department of Justice,[64] these policy considerations are as follows:

(1) Access to confidential information. This may be the most fundamental problem created when government employees who learn of fraud in the course of their official

duties are allowed to file suit under the FCA. By virtue of their official position, government auditors, investigators, attorneys and employees whose duty is to investigate fraud by government contractors, obtain access to confidential, proprietary, and privileged information of companies that serve as government contractors. These government officials also have access to internal governmental information including confidential and non-public records and experts. Access to all of this information is granted only because of the government employees' public position. It is wrong for a government employee to use this access for his personal benefit, rather than to serve the public interest.[65]

(2) Conceal information. An FCA private plaintiff (''relator'') qualifies to serve as a relator only if the information in the relator's FCA lawsuit is non-public.[66] This is a wise limitation for it helps ensure that the relator brings something of value to law enforcement before the government has to share its recovery with the relator. However, when the relator is a government employee who obtained information about fraud in the course of his official duties, this provision of the FCA encourages the government employee to conceal information about fraud from his superiors, co-workers and government prosecutors. By concealing such information, the relator can investigate and develop the case for himself and preserve his eligibility to bring an FCA suit.[67]

(3) Race to the courthouse. The FCA provides that only the first relator to file may bring an FCA lawsuit. Again, this provision is wise policy for this helps ensure that the information the relator brings to the government is not repetitive and justifies DOJ's sharing of its judgment with the relator. This provision also limits, appropriately so, the number of private individuals who share in the government's recovery. However, this provision creates three temptations for the government employee who wants to file his own FCA action.

First, it gives the government employee an incentive to delay the official investigation of the fraud so that the employee has time to prepare and file his FCA law-suit before another potential relator files suit.

Second, this provision encourages the government employee to steer the official investigation in an unproductive or otherwise inefficient direction so as to obfuscate for other potential relators key facts that would enable those persons to file a FCA action and thus beat the government employee in the first-to-file race. For example, if a government employee who is investigating fraud by a government contractor is planning to file his own FCA lawsuit and is concerned that interviewing a particular witness may cause that witness to file his own FCA action, the employee may be tempted to delay the interview until after the employee has filed his own FCA action.

Third, the first-to-file requirement may encourage a government employee to file his FCA lawsuit too early, thereby short-circuiting or derailing other productive avenues of official investigation such as grand jury investigations.

While some of these ''race to the courthouse'' problems may arise with any relator, they are exacerbated when the relator is a government employee because of the unusual access to information and ability to derail the investigation by a government employee who investigates the fraud as part of his duties.

(4) Premature disclosure. Although FCA lawsuits filed by relators are filed under seal, and thus presumably unknown to the outside world including defendants, the government's deadline for reviewing and investigating the case, and deciding whether to join in the case, will begin to run as soon as the relator files suit. In this sense relators drive DOJ's agenda. This is, of course, true whether or not the relator is a government employee. However, because the government employee has access to information not available to other relators, the government employee can file her FCA action even earlier than most other relators.

(5) Damaged credibility. Often the government auditor or agent who investigated fraud by a contractor is a key witness at any civil or criminal trial or administrative hearing. These persons often testify as summary expert witnesses, explaining billing requirements and tracing how the defendant's conduct violated these requirements. When this individual has filed a lawsuit under the FCA in his own name and stands to profit personally by doing so, his credibility as a witness is ruined. This cripples the government's case.[68]

(6) Conflict of interest. There are specific prohibitions against federal employees using "nonpublic government information...to further any private interest,[69] participating in a government matter in which the employee has a financial interest,[70] using public office for private gain,[71] using government property or time for personal purposes,[72] and holding a financial interest that may conflict with the impartial performance of government duties."[73] There are also criminal penalties for federal government employees who participate in matters in which they have financial interests.[74]

When a government employee who obtains information of fraud by a government contractor in the course of the employee's duties, files an FCA action in his own name, all of the above regulations and statutes are violated. They are violated when the government employee reviews documents, interviews witnesses, and discusses strategy and investigative direction with other government employees with expertise in such matters. Access to such information and expertise would not be available to the government employee if anyone knew that the employee was going to use it to reap private gain.[75]

(7) Erosion of public confidence. When potential defendants or witnesses know that the government employee who is investigating fraud may be working for himself and his own profit, they are less likely to come forward to voluntarily cooperate, or to be fully forthcoming.[76]

B. United States ex rel. Bobby Maxwell v. Kerr McGee Oil & Gas Corporation

On June 14, 2004, Bobby Maxwell filed a qui tam action under the civil False Claims Act. He filed his FCA action as a private person even though Maxwell was at the time employed as a senior auditor with the Minerals Management Service ("MMS"),[77] of the United States Department of Interior. In his FCA action, Maxwell alleged that Kerr-McGee underpaid royalties it owed to the federal government pursuant to fifty-seven leases under which Kerr-McGee produced oil offshore in the Gulf of Mexico.[78]

The federal government declined to join in Maxwell's FCA action.[79] Kerr-McGee filed a motion for summary judgment prior to trial arguing that Maxwell, as a government employee who obtained information of Kerr-McGee's alleged fraud in the performance of his official duties, was precluded from bringing suit under the FCA. The court deferred ruling on Kerr-McGee's motion until after trial. A jury trial was held from January 16, 2007, to January 23, 2007. The jury returned a verdict in favor of Maxwell and found damages of $7.5 million.[80] On March 30, 2007, the District Court granted Kerr-McGee's motion for summary judgement, holding that Maxwell did not qualify as an FCA relator.[81]

In granting Kerr-McGee's motion for summary judgment, the District Court noted that it was bound by Tenth Circuit precedent in United States ex rel. Holmes v. Consumer Insurance Group,[82] in which the Tenth Circuit held that ''[a]lthough there may be sound public policy reasons for limiting government employees' ability to file qui tam actions [under the FCA], that is Congress' prerogative, not ours.''[83] For this reason, the District Court reasoned, it could not hold that Maxwell was precluded from filing suit solely by virtue of his status as a government auditor who discovered the alleged fraud in the regular course of his duties.

The District Court noted, however, that Maxwell obtained the information in his FCA suit as part of his official duties. Specifically, the court noted that as the senior auditor charged with determining whether Kerr-McGee's conduct was ''correct...and not fraudulent,'' Maxwell:[84]

- ''...frequently was on site at Kerr-McGee and took documents back to his office to review ''whenever possible.''[85]
- ''...designed the subject audit of Kerr-McGee.''[86]
- requested documents from the field auditors which they obtained pursuant to his requests.[87]
- ''...personally conducted a comparison of the Texon sales price data[88] to fair market value and to the values reported by other companies, and discovered a vast difference.''[89]
- ''...discovered two internal Kerr-McGee memorandums that would later be used to show that Kerr-McGee considered the Texon arrangement to be to its economic advantage.''[90]
- ''...signed the letter from MMS to Kerr-McGee determining that Kerr-McGee had underreported its royalties,[91] and drafted follow-up responses.''[92]

Procedural provisions unique to the FCA require that all relators must qualify as an ''original source'' before they may file an FCA action if the information in the action has been ''publicly disclosed.'' The District Court found that there had been public disclosure of the information in Maxwell's FCA action. The court then held that because of Maxwell's duties as a government employee, he failed to qualify as an ''original source'' and thus, was ineligible to bring his action under the FCA.

The Maxwell case aptly demonstrates the unfairness in allowing a government employee who has access to a company's confidential, proprietary and non-public information and access to governmental records and experts, to use the FCA for his personal gain rather than for the public interest. Every bit of the information Maxwell gathered about Kerr-McGee's alleged fraud was because of his status as a government auditor.

Although the District Court of Colorado ultimately held that Maxwell failed to qualify to bring an FCA lawsuit, and thus reached the ''right'' conclusion, it was precluded, by Tenth Circuit precedent, from doing so on the ground that Maxwell was ineligible to bring suit under the FCA because he was a government employee who obtained the information in his FCA action in the course of his official duties. The court reached the same conclusion but had to labor through the ''public disclosure'' and ''original source'' analysis to do so. It would have made more sense from a policy perspective if the District Court could have dismissed Maxwell's lawsuit at the beginning of the case simply because he was a government employee who obtained information in the course of his duties.

In addition and significantly, the resolution in the Maxwell case was highly factspecific. It was fortuitous that the court was able to resolve the matter on grounds unrelated to Maxwell's status as a government employee. Other instances of government employees who capitalize on their access as government officials for personal gain, may present different facts and may not be subject to dismissal. An amendment to the FCA clarifying that government employees who discover alleged fraud in the course of their official duties should not qualify to bring FCA lawsuits is needed.

III. Proposed Amendment

There is no question that government employees who do not obtain information of fraud in the usual course of their duties should qualify, just like anyone else, to file private actions as relators under the False Claims Act. Thus, for example, a government employee who discovers fraud through friends or social connections in her community should qualify as a relator. The problem arises only when government employees who discover fraud in the course of their governmental duties use that knowledge to profit personally by filing an FCA action. The following proposed amendment to the FCA achieves this balance. Significantly, Congress had recognized and addressed this issue when creating private causes of action for those who know of bank fraud. Congress did so by including the following language in the Financial Institutions Anti-Fraud Enforcement Act of 1990.[93] This statute contains a qui tam provision similar to that in the FCA.

> ''the declaration is filed by a current or former officer or employee of the Federal or State government agency or instrumentality who discovered or gathered the information in the declaration, in whole or in part, while acting within the course of the declarant's government employment.''[94]

Such an amendment to the FCA would address the problem of government employees who seek to profit by filing suit under the FCA using information they obtained in their official capacity. It would also appropriately and fairly allow government employees who obtain information of fraud independently of their official duties, to qualify as relators under the FCA.

The Chairman. The Chair thanks the panel for being with us today.

I want to begin my questions with Mr. Roller. I am quite impressed that North Dakota collects almost $6 for every $1 you spend on audits, increasing Federal royalty collections in North Dakota alone during the period of 1982 through 2001 by $26.6 million. According to your testimony, this experience is multiplied and seen across a spectrum of states and tribes that participate in the MMS delegated audit program.

Mr. Geesey, does the Wyoming program enjoy the same level of success? I am at a loss, therefore, to understand. Well, let me ask you that question first.

Mr. Geesey. Mr. Chairman, I don't have the exact numbers, but I can give you our total collections, which would include Federals, because we do audits for both the Federal and our state side, and right now we stand at about last year we collected about $68 million in additional assessments at a cost of around $3 million.

The Chairman. Let me turn it over to Mr. Costa in my absence.

I am sorry. Let me go ahead and recognize the Ranking Member, Mr. Pearce.

Mr. Pearce. Thank you, Mr. Chairman. I really am sorry I missed the second panel. We had another obligation.

I was looking forward to hearing Mr. Maxwell. He has been honored on the Floor of the House of Representatives as the one guy who accomplished what the Administration would not and could not do prior to that, bringing up possible negligence, lost his job, standing up. It just sounds like the Paul Revere of royalties.

Out in our area we have the Texas Rangers. They enforce the law. It sounds like Mr. Maxwell was really a Texas Ranger. Maybe we could call him a Royalty Ranger at this point for his great service.

I would like to submit a couple of documents. Mr. Maxwell testified that he reported his concerns regarding the Kerr-McGee litigation to both the Office of Enforcement and the IG, so first of all, Mr. Chairman, I would like to have unanimous consent to submit the MMS procedure that requires auditors to report their concerns regarding illegal acts to the Office of Enforcement or IG.

Second, I would like to submit a statement from the MMS stating that Bobby Maxwell never, never bubbled this information up to the IG or the Office of Enforcement and that he then carried it out and used it to apply in court for something that possibly could bring him up to 10 million. I think we have significant concerns about why he did not bubble these things up through the system the way the law requires.

The next thing I would like to have introduced is the opinion of MMS's third party auditors, people who say they are doing a fairly good job, and that is from Thompson, Cobb, Brazilio & Associates.

With your consent, Mr. Chairman, we will enter those into the record.

Mr. Costa [presiding]. With no objection I would be willing to do that.

Mr. Pearce. OK. Thank you.

Mr. Costa. I, as a point, would like to note that with your references Mr. Maxwell was completely exonerated of the charges.

You know, I understand the point you made, but the fact is that the real problem of mismanagement at the Minerals Management Service has existed. Mr. Maxwell took the matter to court.

Mr. Pearce. Mr. Chairman, am I yielding back my time?

Mr. Costa. Let me finish my point, and I will be happy to yield back.

Mr. Pearce. If I could stop the timer. That is the point.

Mr. Costa. No. I am very generous with the time, the gentleman from New Mexico.

Mr. Pearce. OK. Thank you.

Mr. Costa. I have proven that in the previous hearings we have held.

I think it is important, and without objection we will submit the information that you provided. I think it is important to note that he was totally exonerated by the courts, and the documentation is there to prove that.

Without objection, we will submit that as well.

[Note: The information submitted for the record by Mr. Pearce has been retained in the Committee's official files.]

Mr. Costa. I will allow you to continue. The gentleman from New Mexico?

Mr. Pearce. Thank you. I thank the Chairman for his indulgence.

I would point out that the case with Kerr-McGee is still in process, so I am not sure exactly which court exonerated.

Additionally, Mr Chairman, I would point out that the Code, 5 C.F.R. 2635.101(b)(3), 2635.703(a), they prohibit a Federal employee from using nonpublic government information to further their private interests and so I have serious concerns about Mr. Maxwell's legitimacy and in fact his standing whether or not these cross the line at being criminal acts.

Also I would refer to C.F.R. 2635.402, 2635.501, 2635.502, which prohibit a government employee from participating in a government matter in which the employee has a financial interest. Additionally I would point out 2635.101(b)(7), 2635.702, which prohibits a person from using public office for private gain.

I would also point out C.F.R. 2635.704, 2635.705, which prohibits a government employee from using government property or government time for gain, and finally in fact 18 U.S.C. 1905, which in fact it is a crime for government employees to disclose information gathered in the course of their employment.

These are all things which raise grave concerns on the part of the Minority about the testimony and in fact the actions by Mr. Maxwell.

I would go next to a question that—well, I see my time has elapsed. If we get a second round, I will gladly take a second round. I yield back.

Mr. Costa. Thank you very much, the gentleman from New Mexico. As all of us are certainly entitled to our point of view, I am not an attorney and that is a point I will make throughout the 110th Congress as I chair this subcommittee. I certainly appreciate everybody's perception.

Cases that are pending or cases that have been rendered on we can certainly comment here on the Committee, but I would like to focus back on this third panel. I want to apologize for not having listened to your testimony. I had read it last night. Unfortunately, this is one of those days.

Mr. Geesey, I heard one of the previous witnesses I did catch on the second panel talk about his experience on tribal lands both in the sector in his service to government and beyond.

Do you have the experience to comment on communication and cooperation between states and tribes with the Minerals Management Service and give us a flavor from your own vantage point?

Mr. Geesey. Mr. Chairman, no, I don't.

Mr. Costa. OK. Well, as I would not give you legal advice, I am pleased that you are not going to give us advice in an area that you don't feel comfortable with.

Mr. Roller, how about yourself?

Mr. Roller. As for communication between states and tribes within the MMS are you talking?

Mr. Costa. Yes.

Mr. Roller. It mainly is at STRAC meetings. Other than that, I don't know. I guess I don't have any experience to comment other than that.

Mr. Costa. Mr. Lester looks like the gentleman who is nodding his head in great fashion, so maybe you are the best person to ask that question to.

Mr. Lester. I was agreeing with Mr. Roller that the principal communication and interaction between tribes and states is through the State Tribal Royalty Advisory Committee, and that was formed because so many different Federal agencies of the Department of Interior and so many of the stakeholders needed a forum in which to communicate with one another.

In addition, the Department has an intradepartment, an interagency committee, a steering committee on Indian minerals because there is perhaps as many as 12 Interior agencies involved with respect to the development, production and payment to tribes and individuals from Indian lands. The coordination within the Department and with the stakeholders is a serious problem, Mr. Chairman.

Mr. Costa. Are any efforts being done, to your knowledge, to try to address that?

Mr. Lester. I know that there are individuals who are trying to address it within their scope of work, but there is no one in the Department whose job it is to see that the Department fulfills its responsibilities to Indian lands in the mineral development, production and payment arena.

Mr. Costa. Mr. Lester, quickly because I have about a minute and 20 left. Do you think that the current Federal laws on the Indian Mineral Leasing Act and the Mineral Development Act are sufficient today to ensure that Indian energy resources are properly developed and monitored?

Mr. Lester. I think if they were properly implemented, sir, they are except with respect to the interests of the individual Indian landowner whose land is held in trust or is restricted. They have little voice in the process, and that aspect needs to be strengthened.

How to do it? I would turn to the Indian Lands Working Group to provide the information on how best to do that.

Mr. Costa. All right. I thank you for the suggestion.

Mr. Geesey, before my time expires, please explain why the State of Wyoming and private parties deduct fewer costs from royalties on the transportation of coalbed methane than Minerals Management Service does. Would you recommend that Minerals Management Service adopt the Wyoming method?

Mr. Geesey. Mr. Chairman, we have been in discussion with the Minerals Management Service regarding the coalbed methane issue, and it is a simple matter of we think the marketable condition is in a different place than the Minerals Management Service believes it is.

The company thinks it is in another different place. All of that is being litigated, as a matter of fact, because of the complications of the statutes and stuff. I mean, every one of them has a legitimate reason why it ought to be there.

We think ours is the most legitimate, but that is just our feeling.

Mr. Costa. And it raises the most money for the public.

Mr. Geesey. In our mind, and of course for the State of Wyoming, which is who we are trying to look out for.

Mr. Costa. Well, that is your job.

My time has expired. The gentleman from New Mexico?

Mr. Pearce. I thank the Chairman.

Mr. Geesey, tell me a little bit about what you think about STRAC and its relationship with MMS and very shortly. We have a lot of questions. Does it work or not? That is all I want to know.

Mr. Geesey. Mr. Chairman, I don't think it works all that well.

Mr. Pearce. OK.

Mr. Geesey. I think it is a good place where we can talk some issues over. I have had this job now for 12 years, and——

Mr. Pearce. All right. You bet. That is good enough. I appreciate it. I have heard it doesn't work all that well. I mean, we don't cut with a very fine tooth blade up here.

Tell me, Mr. Roller. You get you said $6 for every $1 you spend on audits. When I look at the chart over here, if I can get my staffer to hold that up, I wonder why you get more dollars per audit than some of the other states.

In other words, why don't you just contract out with New Mexico? We do a pretty good job. We collect pretty good revenues. It costs you about seven times what it does in New Mexico. Why is that?

Mr. Roller. I can't answer why New Mexico doesn't contract for our services in North Dakota. I can just tell you those are the statistics. They are MMS statistics. They are not my statistics.

Mr. Pearce. OK. And you all do great work?

Mr. Roller. Yes.

Mr. Pearce. OK. Now, what I am reading down through here, when I read the amount of dollars that you collect for what you spend, I read that North Dakota only collects $60 for every $1 that you spend to audit. Louisiana gets $1,400 for every $1 they spend. New Mexico gets $711 for every $1 they spend. Wyoming, who I have heard criticism of their program today, gets $628.

When I look at the way we overcompensate you and the very paltry returns you get, I am not sure I can verify with dollars on a piece of paper that your service works quite as well as what I have heard in testimony today.

Mr. Geesey, tell me what you feel about the compliance reviews versus straight out audits.

Mr. Geesey. We believe that it is a tool to be used. We have some matching tools and so we used that tool. We were glad that the Minerals Management Service gave us some additional resources to do that.

We have already found some additional funds through that process, and we are glad to do it. It is any kind of enhancement of the audit process itself. I mean, as I stated in my testimony, it doesn't slow down the number of audits we do.

In fact, it focuses us a little bit better than we were able to before because of course it enhances our risk assessment model by going through there and seeing those anomalies that do exist when you look at the entire database using the compliance review process.

Mr. Pearce. OK. Ms. Bucy, in your testimony you gave several things about really allowing Federal employees to take very privileged information and take that information and go into court as private citizens and collect money.

You itemized a series of things. Do you think that that is a circumstance that exists? Are you familiar with Mr. Maxwell's case against Kerr-McGee?

Ms. Bucy. I have read the court opinions, yes.

Mr. Pearce. You have read the court opinions? From your experience as a prosecutor, is that something you would be concerned about?

In other words, that Federal employee who took privileged information and then went into the courts and filed a lawsuit as if he were a private citizen. Did that cross that threshold that you as a prosecutor would have been concerned about?

Ms. Bucy. Yes. Yes, sir, it does in two respects. First of all, I am not so sure that what Mr. Maxwell has identified is fraud. It may just be accounting differences.

I am not so sure that a jury is really capable of telling the difference, so I would feel better if a prosecutor had signed off on it, somebody who knows fraud, and in fact the Department of Justice had the chance to review his case and turned it down because they said it was not fraud; it was just accounting differences.

First of all, I don't know that it is fair to a defendant to have people going around saying that stuff is fraud when it is not. Second, if it is fraud and I as a prosecutor was going to have to prosecute it one of the key witnesses would be the government auditor who reviewed it, and if that person becomes a qui tam relator their credibility is shot and so they have ruined any chance of the government using them as a witness to prosecute that case.

So it is unfair to the government and it is also unfair to the defendant to have government employees pursuing these qui tam cases.

Mr. Pearce. Thank you. I yield back.

Mr. Costa. Mr. Lester, as I read and I believe you testified that you think in terms of as we look at the review that the Committee should provide guidance to the Minerals Management Service as it relates to the degree of monitoring the audits, as well as the royalty payments, so that they are accurate and timely. You also state that there needs to be greater assurances I guess from the Department of Interior that they won't serve as an impediment.

I am wondering on the settlements between the tribes and the private sector parties. Would you please provide examples of how the U.S. Government, which is bound under the Federal law to act as the trustee for the benefit of the tribes, has hindered these dispute settlements?

Mr. Lester. They have hindered in the sense that the data that the tribes depend on often is late and inaccurate, and the tribes' access to company records often is very difficult to obtain.

Often tribes will have to go and dig into reports companies have made to states and compare data on production from the wells because the states also collect taxes from those wells and so they are able to go to the state for information that they can't get from MMS or BLM relative to production.

When the company and the tribe does reach agreement on the settlement of what is owed then MMS or the Federal government requires the tribe to release the government as trustee from any claims; not just about the settlement, but any claims that the tribe may have.

Mr. Costa. That is helpful, Mr. Lester. What I would ask you to do is to submit some suggestions to the Committee on how we might improve this process.

Mr. Lester. Thank you. I would be happy to do that.

Mr. Costa. Yes.

Mr. Geesey, a couple quick questions to you and Mr. Roller before my time expires. In 2006, I understand that the State of Wyoming voiced its opposition to a plan by Minerals Management Service to curtail meetings held between state and tribal auditors.

Despite Wyoming's opposition and others', MMS went ahead with the plan. Can you explain why you opposed this plan?

Mr. Geesey. Mr. Chairman, I am not sure I understand what the question is in terms of a plan.

Mr. Costa. The plan that we were told was meetings between the state and tribal auditors that limited the numbers that Minerals Management Service conducted.

Mr. Geesey. I don't know of any time when we would oppose a meeting with STRAC. We probably think that the number of STRAC meetings is too many, but I don't know that.

Mr. Costa. That is fine. Mr. Geesey and Mr. Roller, according to the whistleblowers, Minerals Management Service moved forward on a plan in 2002 to close hundreds of pending audits so that it could receive a clean audit opinion after four years of unqualified audit opinions.

Were the findings of these audits collected, or did MMS simply walk away from, in your opinion, money that may have been owed to the taxpayer?

Mr. Roller. From North Dakota's perspective, those audits, if they were a North Dakota audit, we followed up on them, but I believe the majority of those audits that were closed were Minerals Management Service audits, which I wouldn't have any information on.

Mr. Costa. Mr. Geesey?

Mr. Geesey. I would echo the same thing.

Mr. Costa. All right. Very good. I will yield back the balance of my time.

This will be the last round. Following the gentleman from New Mexico's questions or comments we will adjourn the hearing, and we want to thank you for your time and your testimony.

Mr. Pearce. Thank you, Mr. Chairman. Again, a great hearing.

Mr. Lester, you stated some concern about the MMS. You talk about some of the tribes doing the monitoring themselves, and I will tell you that I am always talking to the tribes in my district about more self-determination.

Why don't you pick it up and do it? Why don't more tribes just pick this up and do it themselves rather than depending on MMS or whoever?

Mr. Lester. Our ability to enforce contracts and to enforce the law against non-Indians is very limited by Federal law.

We do require the backing, in fact the enforcement authority, from the Federal government as trustee because of that very limited amount of jurisdiction we have over non-Indians.

Mr. Pearce. OK. I appreciate that. That is something that I am sensitive to.

Professor Bucy, if I understand the case, and we are going back to Maxwell one more time. If I understand the case, what happened is that Kerr-McGee were contracting with Texon to sell their production, and Mr. Maxwell said no, that is not appropriate. You could have made more money if you had done this.

Now, each one of us, and I have never sold a house on my own. I have sold lots of houses. We contract with a realtor to get the best we can. We don't know the process. So basically what Kerr-McGee did is contract with the realtor to sell some property. They contracted with Texon to sell their oil. Mr. Maxwell said in his professional opinion they could have made more money. They paid $110 million, and he said now if you had done it this way you could have made $120 million.

Talk a little bit about the legal implications. If this case goes through and Mr. Maxwell is allowed to profit by this private information that he got while he was a public employee, but furthermore if he can direct someone.

Can the IRS come back and tell every American? Can every IRS agent then go into American homes and say you used a realtor to sell your property. If you had sold it yourself, you would have been better off.

Talk a little bit about the risks throughout the system, the shockwaves.

Ms. Bucy. What is at heart in the case against Kerr-McGee is whether or not it got fair market value. You know, that is in the eye of the beholder what is fair market value.

I think it is significant to note that whatever might be the problems with MMS, MMS said this was not fraud. This was a difference in opinion of fair market value. Then it went up to the Department of Justice, and the Department of Justice looked at it for fraud and said no, there is not fraud here. This is a difference in opinion of fair market value.

The ramifications of that judgment standing are significant because not only do you have all of these problems of a government employee having access to confidential proprietary information, but using it to say that there is fraud when it is just a difference of opinion is going to completely undercut the effectiveness of the False Claims Act.

Mr. Pearce. And that undercutting would then generate these perverse incentives that cause good government employees to act in a parasitic way. I think that is a very disconcerting possibility.

Would you outline the legislative fix that you had suggested in your testimony just briefly? How would it be fair to government employees and yet be fair to the people and not incentivize people to act like parasites?

Ms. Bucy. The language would only prevent government employees from serving as qui tam relators if they got that information acting in the course of their government employment, so obviously it would apply to attorneys, auditors, anybody who is getting confidential proprietary information in the course of their employment from turning around and using it for private gain.

It would still allow, and I think appropriately so, government employees to be qui tam relators if they get the information totally aside from their government duties.

Mr. Pearce. So it would give fairness, but also not incentivize with perverse incentives this parasitic behavior?

Ms. Bucy. Exactly. Exactly.

Mr. Pearce. OK.

Ms. Bucy. I should note this is something that Congress passed. I mean, that language is not my own. It is something that Congress has already passed, but with regard to qui tam provisions in bank fraud.

Mr. Pearce. In your opinion, was there court shopping going on here? In other words, it should have been filed in the Fifth, but it was instead in the Tenth.

Ms. Bucy. Absolutely. Well, forum shopping in the sense that that is what plaintiffs' attorneys do and are allowed to do, properly so, according to the courts.

The contract was in the Gulf of Mexico. That is the Fifth Circuit. The Tenth Circuit is where it was filed, and that is because the Tenth Circuit allows these parasitic lawsuits by the government employees to go forward whereas other Circuits would not.

The Fifth has not yet ruled on it, but the Tenth has, and it is favorable to the government employees.

Mr. Pearce. Thank you.

Ms. Bucy. The First Circuit and the Ninth Circuit are the ones that have clearly ruled that these are inappropriate lawsuits.

Mr. Pearce. Thank you, Mr. Chairman. A great hearing. I appreciate it.'

Mr. Costa. All right. I want to thank those who testified on the third panel, the second panel and the first panel on behalf of Chairman Rahall and all those who participated today.

I thank the Members of the Committee. This hearing is now adjourned.

[Whereupon, at 2:28 p.m. the Committee was adjourned.]

[Additional material submitted for the record follows:]

[A statement submitted for the record by Erich G. Pica, Director of Domestic Programs, Friends of the Earth, follows:]

Statement Submitted for the Record by Erich G. Pica, Director of Domestic Programs, Friends of the Earth

On behalf of Friends of the Earth, I would like to thank the Committee on Natural Resources for the opportunity to offer testimony for the record on the Department of the Interior's collection and management of oil and gas royalties. Friends of the Earth is a national non-profit environmental advocacy organization and is a member of Friends of the Earth International, the world's largest grassroots environmental federation with more than one million members in 71 countries.

Last year the New York Times began publishing a series of investigative articles that exposed gross mismanagement of the Department's royalty program which could cost taxpayers billions. These articles spurred welcomed congressional oversight and attention to the issue of royalties paid by oil and gas companies for the privilege of drilling on federal lands and waters. The articles and subsequent congressional investigations have uncovered a pervasive culture of ineptitude that has put at risk tens of billions of royalty dollars. The failure to collect royalties means that millions of acres of public lands and waters are being put at environmental risk without fair taxpayer return. These royalties provide needed funding to the Land and Water Conservation Fund, the Historic Preservation Trust Fund, the oil-producing states and the general treasury.

The Interior Department itself first revealed the problem in last year's budget, which noted that ''royalty relief'' would allow companies to avoid paying royalties on more than $65 billion worth of revenues from oil and gas drilled in the deep waters of the Gulf of Mexico over the next five years, costing the federal government approximately $9.5 billion over that period[95]. It was later discovered that a large share of the losses resulted from the failure to include price thresholds capping royalty relief in leases issued in 1998 and 1999.[96] According to a draft report by the Government Accountability Office, losses to the treasury over 25 years could reach a staggering $20 billion due to a combination of the missing price thresholds and a recent federal court decision that changed the methodology by which royalty relief is calculated. If the oil industry is successful in a recent legal challenge, these losses could balloon to $80 billion over the same period.[97]

Unfortunately, the missing price thresholds are only the tip of the iceberg. Following the discovery the erroneous 1998 and 1999 leases, media, congressional, and departmental investigations and whistleblower actions have highlighted the failure of the Minerals

Management Service (MMS) to audit royalty payments or to seek payment of underpaid royalties and interest on the royalties. Now that MMS's failures are well-documented, it's time for Congress to insist on badly-needed reforms to the agency. Friends of the Earth proposes the following reforms that we believe will make oil and gas companies more accounted to the American taxpayer.

Establish Independent Auditing

MMS has lost the confidence of Friends of the Earth, Members of Congress and the American public, and cannot be trusted to fully and fairly collect royalties from oil and gas companies on behalf of the American public. Indeed, MMS is plagued by a culture of ineptitude that makes it unfit to manage the nation's royalty collection program. Friends of the Earth recommends that MMS's royalty collection and auditing functions be either put into receivership with an outside auditing firm, or placed under the direct guidance of the Department's Inspector General. If these recommendations appear insufficient, Friends of the Earth would recommend evaluating the potential of housing the royalty collection and enforcement with the Internal Revenue Service.

Increase Transparency

As investigations by the Department's Inspector General and the Government Accountability Office have progressed, one common theme continues to reappear: the lack of readily available and verifiable information regarding oil and gas royalty payments. As a founding member of the Publish What You Pay Coalition-US, a coalition of more than 300 non-governmental government organizations worldwide helping citizens of resource -rich countries hold their governments accountable for the management of oil revenues, Friends of the Earth is keenly interested in the full, timely and verifiable disclosure of royalty payments made to the federal government. Royalty payments, contracts between the federal government and companies, must be accessible and understood by the general public. In addition, the public must be notified about when a company is being audited, the results of the audits, and any penalties/rewards levied after the completion of an audit.

Increase the Royalty Rates

Currently, oil and gas companies typically pay a 12 to 16 percent royalty on oil and gas they extract from federally owned waters and lands. According to the New York Times (Incentives on Oil Barely Help U.S., Study Suggests 4), the United States imposes significantly lower royalty and tax rates on oil and gas companies than other countries. According to the article:

In the United States, the federal government's take—royalties as well as corporate taxes—is about 40 percent of revenue from oil and gas produced on federal property, according to Van Meurs Associates, an industry consulting firm that compares the taxes of all oil-producing countries.

By contrast, according to Van Meurs, the worldwide average ''government take'' is about 60 to 65 percent. And that figure, of course, excludes countries that do not allow any private ownership in oil production.

With annual profits of the big five oil and gas companies exceeding more than $110 billion last year, it is time that Congress reevaluates the royalties in place for current and future production.

In addition to the low tax and royalty rates, the impact of royalty relief to spur domestic production is dubious. An MMS report entitled ''Effects of Royalty Incentives for Gulf of Mexico Oil and Gas Leases''[99] demonstrates this point. Table 5-9 of Volume 1 reviews various royalty relief scenarios and estimates their impact over the next 40 years. Under the current royalty relief regime, MMS is estimating that 57,281 mmBoE in new reserves will be discovered, they estimate that 56,644 mmBoE will be discovered without royalty relief. This is a difference of approximately 1.1 percent. This additional 1.1 percent in discovery will cost taxpayers approximately $48 billion in royalty revenues, the difference between the total royalty revenue from the current royalty regime and no relief.

Table 5-9. Effects of Alternative Royalty Scenario on Activities at All Fields.

Royalty Scenario	Water Depth	Oil Price ($/bbl)	Gas Price ($/mcf)	Reserves Discovered (mmBOE)	Total Gas Production (Bcf)	Total Oil Production (mmBbl)
DWRR - Field	All	46	6.96	57,510	173,134	33,092
DWRR - Lease	All	46	6.96	58,062	174,240	33,382
Current	All	46	6.96	57,281	172,734	32,992
No Relief	All	46	6.96	56,644	171,638	32,699

Table 5-9 (continued)

Royalty Scenario	Water Depth	Oil Price ($/bbl)	Gas Price ($/mcf)	Royalty-Free Gas Production (Bcf)	Royalty-Free Oil Production (mmBbl)	Royalty-Paying Gas Production (Bcf)	Royalty-Paying Oil Production (mmBbl)
DWRR - Field	All	46	6.96	29,801	7,113	143,334	25,979
DWRR - Lease	All	46	6.96	47,944	11,747	126,296	21,635
Current	All	46	6.96	21,233	4,988	151,501	28,003
No Relief	All	46	6.96	11,442	2,397	160,196	30,302

Table 5-9 (continued)

Royalty Scenario	Water Depth	Oil Price ($/bbl)	Gas Price ($/mcf)	Total Royalty Revenue (mm)	Present Value Royalty Revenue (mm)	Present Value of Lease Bonus and Rental Revenue (mm)	Total Present Value of Fiscal Variables (mm)
DWRR - Field	All	46	6.96	$540,789	$90,172	$5,300	$95,472
DWRR - Lease	All	46	6.96	$454,533	$85,112	$6,099	$91,212
Current	All	46	6.96	$577,720	$92,864	$4,290	$97,154
No Relief	All	46	6.96	$625,735	$97,367	$3,823	$101,190

Note: Values in $2003. Present value at 12%. Reserves are ultimate (grown) amounts.

Eliminate Oil Royalty Relief Programs

For more than a decade, Friends of the Earth has opposed various schemes to provide oil and gas companies with royalty relief. We have consistently opposed bills such as the Deepwater Royalty Relief Act of 1996, royalty relief provisions in the Energy Policy Act of 2005, as well as other schemes to help oil companies avoid paying royalties. Particularly at a time of record oil and gas prices, it is clear that companies do not need these incentives. Friends of the Earth was pleased that H.R. 6, the CLEAN Energy Act of 2007, repealed royalty relief provisions enacted in the 2005 energy bill. We are hopeful that the Senate will follow suit, and we urge Congress to repeal additional royalty relief provisions such as sections 343 and 353 of the Energy Policy Act of 2005 dealing with royalty relief for marginal well production and methane gas hydrates.

Collect Royalties from the 1998 and 1999 Leases

Friends of the Earth remains concerned that MMS is not doing enough to recapture the royalty revenue lost as a result of the 1998 and 1999 leases in the Gulf of Mexico. We remain disheartened by statements made by Interior officials regarding their ability to renegotiate the existing leases, as well as the terms of leases that have been renegotiated to date. Members of this Committee played a lead role in passing legislation last year that provided a strong incentive to companies benefiting from unlimited royalty relief to sit down at the negotiating table and agree to begin paying royalties. As Congress sought to strengthen the Department of the Interior's hand with these measures, administration officials such as MMS Director Burton undermined them every step of the way by opposing the congressional actions and publicly stating that ''I don't like to say 'negotiate' because I really don't have anything to trade.''[100]

The terms of the renegotiated leases that six companies have now signed are highly unfavorable to the government and illustrate the result of MMS's weak attempts to address the problem. Among other things, the leases fail to recoup royalties on past production that occurred on the 1998 and 1999 leases prior to October, 2006—nearly $1 billion in lost royalties. The leases also contain a variety of escape hatches for the companies, allowing them to terminate the leases if the Department reaches ''more favorable'' terms with another company in the future or loses a lawsuit filed by Kerr Mc-Gee challenging its authority to collect royalties, or if Congress enacts legislation intended to address the faulty leases. Rep. Hinchey has drawn attention to these flaws, and we share his concern.

In addition to the favorable renegotiated contracts, the Department of the Interior is now asking for authority from Congress to offer lease extensions as a means to entice companies to renegotiate. In testimony before the Senate Energy and Natural Resources Committee, the Assistant Secretary for Lands and Management, Stephen Allred, requested additional authority from Congress to allow the Department of the Interior to extend leases for companies that choose to renegotiate their 1998 and 1999 leases. After opposing legislation to strengthen their negotiating stance, and making public statements that weakened it, MMS is now asking Congress for authority to give yet another giveaway to the oil and gas industry. Needless to say, Friends of the Earth would oppose these incentives.

Conclusion

The oil royalty collections system in the United States is broken. The Department of the Interior and the MMS have lost the ability to enforce the laws passed by Congress and protect the environment and taxpayer interests. I thank the Committee on Natural Resources for allowing Friends of the Earth to offer testimony, and hope our recommendations will be considered in further debate.

End Notes

[1] Minerals Management Service's Compliance Review Process, Department of the Interior Office of the Inspector General, Report No. C-IN-MMS-0006-2006 (Washington, D.C.: Dec. 2006).
[2] GAO, Suggested Areas for Oversight for the 110th Congress, GAO-07-235R (Washington, D.C.: Nov. 17, 2006).

[3] GAO, Royalty Revenues: Total Revenues Have Not Increased at the Same Pace as Rising Natural Gas Prices due to Decreasing Production Sold, GAO-06-786R (Washington, D.C.: June 21, 2006).

[4] The remaining $1.9 billion consist of other revenues received from rent payments and bonuses paid by companies for successful bids on leases.

[5] Geothermal energy is literally the heat of the earth. This heat is abnormally high where hot and molten rocks exist at shallow depths below the earth's surface. Water, brines, and steam circulating within these hot rocks are collectively referred to as geothermal resources.

[6] 30 U.S.C. § 191(a). The State of Alaska is an exception to this provision, receiving 90 percent.

[7] GAO, Mineral Revenues: A More Systematic Evaluation of the Royalty-in-Kind Pilots is Needed, GAO-03-296 (Washington, D.C.: Jan. 9, 2003) and GAO, Mineral Revenues: Cost and Revenue Information Needed to Compare Different Approaches for Collecting Federal Oil and Gas Royalties, GAO-04-448 (Washington, D.C.: Apr. 16, 2004).

[8] GAO, Renewable Energy: Increased Geothermal Development Will Depend on Overcoming Many Challenges, GAO-06-629 (Washington, D.C.: May 24, 2006), 34-38.

[9] EIA U.S. Imports by Country of Origin, 12-21-2006.

[10] Bainbridge Professor of Law, University of Alabama School of Law (law faculty at UA School of Law, 1987 to present); Assistant United States Attorney, E.D. MD., 1980-1987; Law Clerk, The Honorable Theordore McMillian, United States Court of Appeals for the Eighth Circuit, 1978-1979; J.D., Washington University, 1978.

I am most appreciative to Daniel Everett and to the following students in my course on White Collar Practice who helped me prepare this testimony: Bob Elliott, Prim Formby, Clay Gunn, Matthew Harris, Emily Hines, Glenn Jones, Emily Kornegay, Mike Kuffner, Will McComb, Monique Nelson, Robert Pitman, Oscar Price, Scott Sanders, Sirena Saunders, Matt Shelby, Joseph Sherman, Harrison Smith, Derrick Williams.

I am also greatly appreciative to Creighton Miller and Penny Gibson at the University of Alabama School of Law Library, and to Erica Nicholson, my assistant.

[11] 31 U.S.C. § 3729 et seq.

[12] Case 1:04-CV-01224-PSF-CBS (D.Colo., March 30, 2007).

[13] 31 U.S.C. § 3729 et. seq.

[14] Act of March 2, 1863 at ch. 67, 12 Stat. 696-98.

[15] Rev. Stat. 3490-94 and 5438 (1875); 89 Cong. Rec. S7606 (Sept. 17, 1943); Pub. L. 99-562, 100 Stat. 3153 (1986); Pub. L. 103-272, 108 Stat. 1362 (1994).

[16] Pub. L. 99-562, 100 Stat. 3153 (1986).

[17] Hearings Before House Comm. On Judiciary, Subcomm. on Immigration and Claims, 105thCong., 2d Sess.14 (1998) [hereinafter Subcomm. on Claims Hearings] (Testimony of Donald K. Stern, U.S. Attorney, Dist. Mass. and Chair, Attorney General's Advisory Comm., U.S. Dept. of Justice).

For example, in FY 2000 the United States collected $1.5 billion in civil fraud recoveries, most of which, $1.2 billion, was collected through a private justice action, the qui tam provisions of the False Claims Act (FCA). Press Release, United States Department of Justice, November 2, 2000 at www.USDOJ.Gov, 21 TAF Qtrly. Rev. 18 (Jan. 2001) [hereinafter Press Release, DOJ, Nov. 3, 2000].

As one Department of Justice official explained in 1996: "The recovery of over $1 billion demonstrates that the public-private partnership encouraged by the Statute [the FCA] works and is an effective tool in our continuing fight against fraudulent use of public funds." Taxpayers Against Fraud, The 1986 False Claims Act Amendments, Tenth Anniversary Report 15 (1986) (quoting Frank W. Hunger, Assistant Attorney General, Civil Division, U.S. Dept. of Justice).

See also, Hearings Before House Comm. On Judiciary, Subcomm. on Immigration and Claims, 105thCong., 2d Sess.14 (1998) [hereinafter Subcomm. on Claims Hearings] (Testimony of Donald K. Stern, U.S. Attorney, Dist. Mass. and Chair, Attorney General's Advisory Comm., DOJ) ("[T]he False Claims Act...has been the Department's primary civil enforcement tool to combat fraud..."); Id. at15 (Testimony of Lewis Morris, Assistant Inspector General, Dept. of HHS) ("The False Claims Act has been an essential tool to protect the integrity of the Medicare program." "To achieve this goal...of "zero tolerance" of Medicare fraud and abuse...the Government relies on a number of enforcement options—criminal, civil, and administrative, as well as educational outreach efforts. Chief among the enforcement tools has been the False Claims Act."); Id. at 25. (Testimony of Robert A. Berenson, Director for Health Care Plans and Provides Administration, Health Care Financing Administration, Dept of HHS)("[T]he False Claims Act is an important tool for...law enforcement...to pursue fraud and abuse."); Id. at 63. (Statement of Ruth Blacker, National Legislative Counsel, American Association of Retired Persons) ("...Congress in recent years [has] expand[ed] statutory authority and income resources to deal with the problem [of heath care fraud and abuse]. However, none of these things are likely to play a more important role in recovering improper payments to in acting as a deterrent than the False Claims Act. Use of the FCA by Federal authorities has become an important tool for fighting fraud and abuse in many programs, including the Medicare program.")

[18] Bucy, Private Justice, 75 S.C.L. Rev. 1, 13-54 (2002); Pamela H. Bucy, Information as a Commodity in the Regulatory World, 39 Hous. L. Rev. 905, 909-917 (2002) [hereinafter Bucy, Information as a Commodity]; J.

Randy Beck, The False Claims Act and the English Eradication of Qui Tam Legislation, 78 N.C.L.Rev.539 (2000) [hereinafter Beck, English Eradication]; Note, The History and Developments of Qui Tam, 1972 Wash U. L. Q. 81 [hereinafter History and Developments].

[19] See, e.g., The Economic Communications Privacy Act, 18 U.S.C. § 2520 et seq.(2000 & Supp. 2001); American Disabilities Act, 42 U.S.C. § 12188 (a)(1) (1995); The Civil Rights Act of 1964, 42 U.S.C. 1981 et seq (1994) (implied under Title VI, Guardian Assn. v. Civil Service Comm'n of the City of New York, 463 U.S. 582, 594 (1983) and Title IX, Cannon v. University of Chicago, 441 U.S. 677, 717 (1970).

[20] 31 U.S.C. § 3730(b)(1).

[21] Id. at § 3730(b)(2).

[22] 31 U.S.C. § 3730(b)(2)(2002).

[23] For other examples of FCA qui tam cases where the relator and relator's counsel assumed large amounts of responsibility for the preparation of the case, see United States ex rel. Alderson v. Quorum Health Group, Inc., 171 F. Supp.2d 1323 (M.D. Fla. 2001); United States ex rel. Merena v. SmithKline Beecham Corp., 114 F. Supp. 2d 352 (E.D.Pa. 2000)(facts more fully discussed in Merena, 52 F.Supp. 2d 420 (E.D.Pa. 1998) rev'd 205 F.3d 97 (3rd Cir. 2000))

[24] Id. at § 3730(c)(3)(2002).

[25] Id. at § 3730(c)(3)(2002).

[26] 31 U.S.C. § 3730(c)(2)(A) and (B).

[27] 31 U.S.C. § 3730(c)(2)(C).

[28] Id. at § 3730(c)(5).

[29] 31 U.S.C. § 3730(c)(3).

[30] Recent relators' awards include $95 million, $44.8 million, $28.9 million, and $18.1 million. 21 TAF Q. Rev. 20-21 (Jan. 2001).

[31] 171 F. Supp.2d 1323 (M.D. Fla. 2001).

[32] Id. at 1325.

[33] Form HCFA 2552, Cost Reports for Hospitals. Cost reports are lengthy and complex, consisting of hundreds of worksheets and requiring detailed information about the facility, its staff and operation. Providers are required to allocate various costs, including capital expenditures, medical education costs, travel, malpractice insurance premiums and payments, and every type of patient care costs to various centers, designated by whether the patient was a Medicare patient and whether the expense is properly reimbursable to the Medicare program. Robert Fabrikant, Paul E. Kalb, mark D. Hopson & Pamela H. Bucy, Health Care Fraud, Enforcement and Compliance § 2.02[4] (LJSP 2003) [hereinafter Fabrikant et al, Health Care Fraud].

[34] Id.

[35] 171 F. Supp. 2d at 1325.

[36] Id.

[37] Id.

[38] Id. at 1325.

[39] Kurt Eichenwald, He Blew the Whistle, and Health Giants Quaked, N.Y.Times, Oct. 18, 1998, at Sec. 3, Page 1.

[40] 171 F. Supp. 2d at 1330.

[41] Approximately one year after filing his pro se qui tam complaint, Alderson retained a law firm that specialized in health care law to handle his qui tam case. 171 F. Supp. 2d at1325. In 1995, Alderson changed to a law firm that specialized in FCA qui tam cases. Id. at 1327. This firm represented Alderson until the case was resolved.

[42] 171 F. Supp. 2d at 1325-1329.

[43] Id.

[44] DOJ attorneys believed the fraud to be $10 million or less, too low to consider. Id. at 1325-1331.

[45] Id.

[46] Id. at 1325.

[47] Id.

[48] Id.

[49] Id. at 1326.

[50] Id.

[51] Id.

[52] Id. at 1339.

[53] Panel: FCA Enforcement in the Post-Stevens World, ABA Nat'l Inst. On The Civil False Claims Act and Qui Tam Enforcement, Nov., 2000 (Discussion with Michael Hertz, Director, Commercial Litigation Branch, Civil Division, U.S. Dept. of Justice).

[54] 171 F. Supp. 2d at 1331-34.

[55] According to the court, "Only [Alderson's] dogged resolution, eventually supported by competent professionals and an occasionally reluctant government, resulted in the millions now available for distribution." Id. at 1337.

[56] Id. at 1337-1338.

[57] According to the court, "The record establishes that Alderson's counsel contributed significantly (in both quality and quantity) and at certain moments crucially to this case. That contribution deserves manifest and telling

weight in determining the proper relator's award.'' Id. at 1335. The award of attorneys fees and costs was in addition to the contingency portion of his award that Alderson agreed to pay to his attorneys. Id.

[58] 31 U.S.C. § 3730(d)(1)(2002).

[59] Id. at 1338.

[60] view is more nuanced that simply holding, as the court did, that the FCA as currently constructed does not bar all government employees from bringing FCA actions. The Court went on to hold that the relator in the case before it, an attorney with the United States Department of Defense whose job was to review defense contracts for fraud, was barred from bringing an FCA action. In so holding, the court reasoned that the relator's ''responsibility, a condition of his employment [was] to uncover fraud. The fruits of his effort belong to his employer—the government.'' LeBlanc, 913 F.2d at 20.

[61] United States ex rel. Fine v. Chevron, 72 F.3d 740, 745 (9th Cir. 1995); United States ex rel. LeBlanc V. Raytheon Co., 913 F.2d 17, 20 (1st Cir. 1990).

[62] United States ex rel. Holmes v. Consumer Ins. Group, 318 F.3d 1199, 1214 (10th Cir. 2003); cf. United States ex rel. Williams v. NEC Corp., 931 F.2d 1493, 1501 (11th Cir. 1991).

[63] Holmes, 318 F.3d at 1214. See also, Williams, 931 F.2d at 1503. (''We recognize that the concerns articulated by the United States may be legitimate ones, and that the application of the False Claims Act since its 1986 amendment may have revealed difficulties in the administration of qui tam suits, particularly those brought by government employees. Notwithstanding this recognition, however, we are charged only with interpreting the statute before us and not with amending it to eliminate administrative difficulties.'')

[64] As described by the Department of Justice, Holmes, 318 F.3d at 1212.

[65] See discussion of this problem at Holmes, 318 F.3d at 1217 (Tacha, dissenting); Fine, 72 F.3d at 745 (citing with approval arguments made by the government).

[66] Section 3730(e)(4) of the FCA specifies the type of public information that bars a private plaintiff from bringing suit. ''Public'' for these purposes is only: ''public disclosure of allegations or transactions in a criminal, civil, or administrative hearing, in a congressional, administrative of Government Accounting office report, hearing, audit, or investigation, or from the news media...'' If the information is ''public,'' the private plaintiff must prove that he is the ''original source'' of the information. Id.

[67] See Fine, 72 F.3d at 745 (citing with approval this argument made by the government).

[68] See Fine, 72 F.3d at 745 (citing with approval this argument made by the government).

[69] 5 C.F.R. §§ 2635.101(b)(3), 2635.703(a).

[70] 5 C.F.R. §§ 2635.402, 2635.501, 2635.502

[71] 5 C.F.R. §§ 2635.101(b)(7), 2635.702

[72] 5 C.F.R. §§ 2635.704, 2635.705.

[73] 5 C.F.R. § 2635.403; Holmes, 318 F.3d at 1225 (Tacha, dissenting); cf. Fine, 72 F.3d at 746 (Kozinski, concurring) and at 747 (Trott, concurring).

[74] 18 U.S.C. § 208.

[75] For discussions of this conflict of interest issue see Holmes, 318 F.3d at 1212 (citing arguments made by the government); 318 F.3d at 1224-1225 (Tacha, dissenting); Williams, 931 F.2d at 1503 (citing arguments made by the government).

[76] See Fine, 72 F.3d at 745 (citing with approval this argument made by the government).

[77] Case 1:04-CV-01224-PSF-CBS (Order of Dismissal for Lack of Subject Matter Jurisdiction) at p. 4 (D.Colo., March 30, 2007).

[78] Id. at 2 and 5.

[79] Id. at 11.

[80] Id. at 1 and 6. This amount would be trebled according to the FCA.

[81] Id. at 28.

[82] 318 F.3d 1199 (10th Cir. 2003).

[83] Id. at 10 citing Holmes, 318 F.3d at 1214.

[84] Id. at 2.

[85] Id. at 3.

[86] Id. at 9.

[87] Id. at 9.

[88] Texon was a subcontractor Kerr-McGee dealt with. Id. at 3.

[89] Id. at 9.

[90] Id. at 10.

[91] Id. at 3.

[92] Id. at 4.

[93] PL 101-647 (Nov. 29, 1990).

[94] 12 U.S.C. § 4204(a)(1).

[95] Mineral Management Service. Fiscal Year 2007 Budget Justifications and Performance Information, page 169. http://www.mms.gov/PDFs/2007Budget/FY2007BudgetJustification.pdf

[96] Andrews, Edmund L. ''U.S. Has Royalty Plan to Give Windfall to Oil Companies.'' New York Times. Feb. 14, 2006. http://www.nytimes.com/2006/02/14/business/14oil.html?pagewanted=1&ei=5088&en=87dc413fa6add582&ex=1297573200&partner=rssnyt&emc=r

[97] Government Accountability Office Draft Briefing on Oil and Gas Royalties. March 27, 2006. http://www.nytimes.com/ packages/pdf/business/29lease.pdf

[98] Andrews, Edmund L. ''Incentives on Oil Barely Help U.S., Study Suggests.'' New York Times. December 22, 2006. http://www.nytimes.com/2006/12/22/washington/22royalty.html?ei=5088&en=3c13b8d3062224f4&ex=1324443600&adxnnl=1&partner=rssnyt&emc=rss&adxnnlx=1175098448-K1vw88CiZYOjyxBNKgCuwg

[99] Mineral Management Service Economic Division. ''Effects of Royalty Incentives for Gulf of Mexico Oil and Gas Leases.'' Volume 1. OCS Study MMS 2004-077.

[100] Andrews , Edmund L. ''Interior Official Says She Will Not Try to Recoup Lease Money.'' New York Times. September 22, 2006. ''http://www.nytimes.com/2006/09/22/business/22oil.html?ei=5090&en=cf1f7de8c9d7d739&ex=1316577600&partner=rssuserland&emc=rss&pagewanted=print

In: Royalties at Risk in the Development...
Editor: Thomas Marinelli

ISBN: 978-1-60692-837-0

Chapter 4

THE FEDERAL ROYALTY AND TAX TREATMENT OF THE HARDROCK MINERAL INDUSTRY: AN ECONOMIC ANALYSIS*

Salvatore Lazzari

SUMMARY

Under current law, the hardrock mineral industry pays no royalty to the federal government for the privilege of extracting resources from federal lands. This differs from the federal policy toward the coal and oil/gas industries, the policy of State governments, and the leasing arrangements in the private sector, which often require bonus bids and an ad-valorem royalty on the value of the resulting output. Hardrock mining on acquired federal lands pays a 5% royalty.

The current federal policy toward hardrock minerals is inconsistent with the fundamental market principle that a royalty is a factor payment, part of the rent paid, or the return, to land as both a marketable capital asset and input to production. In general, the free development of federal mining land will result in more public land developed and more minerals produced than is economically efficient. Another implication is that any economic rents, i.e., excess profits to a resource owner above the level required to produce or supply the resource in the long run, would accrue to private rather than public beneficiaries. The free development regime for hardrock minerals on federal lands was created to stimulate economic development of the west and has more recently been sustained to protect the viability of the United States hardrock mineral industry and to prevent negative economic impacts on western communities built around that industry. Introducing a royalty payment system might have an adverse economic effect on hardrock mineral producers but it would also tend to increase output in the rest of the economy and promote a more efficient use of national resources. The hardrock mining industry generally has, over the last four years, been booming, and any adverse industry effects would tend to be mitigated.

* This is an edited, reformatted and augmented version of CRS Report for Congress publication, Publication RL34268, dated November 30, 2007.

The appropriate royalty system, according to economic principles, is the advalorem royalty based on the market value of the mineral upon extraction, adjusted for any externality-related taxes. Using "gross income for depletion purposes" is conceptually the same as market value, although it has practical advantages since that is the basis for a producer's computation of its percentage depletion allowance for tax purposes. Using net smelter return — the basis for most private royalty contracts — is conceptually the same as gross income, although there may be differences due to deductions for costs, and would require a new administrative apparatus. Using the producer's net profits as the basis would render the royalty an income tax, which would be inconsistent with mainstream economic principles that a royalty is part of a factor payment, and would likely reduce, and at times totally eliminate, royalty payments. The U.S. hardrock minerals industry pays income taxes, including the alternative minimum tax, and, in addition, is assessed a variety of claims and patent fees. In addition, hardrock mining firms qualify for some special tax benefits or subsidies: expensing (i.e., a current deduction) of exploration and development costs; the percentage depletion allowance, at rates ranging from 14% to 22% of gross income; and a deduction for mine closing and reclamation costs. The special mining tax breaks are sufficient to lower the effective marginal tax rate slightly below that for other industries. These tax rates, however, are currently much higher than the historical rates, which were either close to zero or negative.

Since 1872, hardrock mining on federal lands has been governed by the General Mining Law, which allows virtually free access to open public domain lands for mineral prospecting, staking claims, exploration and development, and extraction without payment of a royalty.[1] And until annual moratoria were imposed by Congress starting in 1994, claimants on hardrock mineral lands could take title to both the mineral rights and the land, after paying to the U.S. Department of Interior (DOI) the required fee or either $2.50 or $5.00 per acre, and after demonstrating diligent development of the mining claim.

Proposals to reform the Mining Law have been made almost from the time it was signed by President Grant, and throughout its 135 year history. Many Presidents — Presidents Roosevelt, Nixon, Carter, and Clinton — have endeavored to reform the law, as have many in Congress. One of the contentious issues has been that the law makes no provision for the payment of a royalty on the production of the hardrock minerals from federal lands. The royalty-free treatment of the hardrock mineral industry is unlike mining arrangements (leases) on private and state lands, mining of the leasable energy minerals (coal, oil, and gas) on federal lands, and even unlike hardrock minerals on *acquired* (as contrasted with public) federal lands, all of which provide for a royalty payment to the landowner.[2]

In recent years, the Congress has repeatedly considered, but not passed, reform of the Mining Law, including, in part, to require the payment of royalties. In 1990, committees held several hearings on proposed Mining Law reform, e.g., on S. 1126, (101st Congress) which among other things would have imposed an 8% ad-valorem royalty. In 1993, President Clinton's comprehensive economic proposal, part of the State of the Union Address, included claims fees and a 12.5% ad-valorem royalty.[3] Several bills were introduced in the 103rd Congress to impose a royalty (H.R. 322, H.R. 1708, S. 257, S. 375, and S. 775) and hearings were held in 1993.[4] S. 775 passed the Senate in May 1993 and was incorporated into H.R. 322 which was approved by the House in November 1993. H.R. 322 went to conference in 1994 but was not approved reportedly because of opposition from Western lawmakers, mainly over the royalty issue.[5] In 1995, there was an unsuccessful attempt to incorporate a royalty as part of the FY1996 budget. In its FY1998 budget proposal, and again in FY2001, the Clinton Administration proposed a 5% "net smelter" royalty but, like previous efforts, this failed.[6]

At this writing, the major reform bill in the House — there is currently no Senate bill — is H.R. 2262, which, among other things, would impose an 8% advalorem royalty based on the value of the minerals as defined under the gross income definition of the income tax code (essentially the mining firm's sales revenue) and use the proceeds for the cleanup of abandoned sites. The 8% royalty would apply to new production. Production from existing mining operations would pay a 4% royalty. On October 2, 2007, the Subcommittee on Energy and Minerals of the House Natural Resources Committee held a hearing on this bill. A substitute bill was marked up and approved by the full committee (the House Natural Resources Committee) on October 23.[7] The House approved this committee bill, with minor amendments, on November 1 by a vote of 244 to 166. A summary of the provisions of the Houseapproved version of H.R. 2262 is described in the Appendix to this report.

This report analyzes the economic issues underlying proposals to impose a royalty on hardrock (locatable) minerals on public domain lands.[8] It also discusses the federal tax treatment of the hardrock mineral industry. In particular, the report addresses the following questions:

- What exactly is a royalty, and how does a royalty differ from other rental payments or, for that matter, taxes?
- Were a royalty imposed, what would be the best way to structure such a royalty? Should the royalty be an ad-valorem type (i.e., one based on the value or price of the mineral), a fixed unit based royalty (i.e., one based only on mineral output), or a net profits royalty (i.e., one based on the net income or profit of the mining company)?
- If there is to be an ad-valorem royalty based on price or value, at what stage in the mineral producing process should value (or price) be measured? Should it be based on gross income (value of the mineral at the mine) or net smelter return (the value at the smelter less certain deductions)? When does a mining company produce a saleable or marketable product whose value can be measured? What deductions, if any, should be allowed from that price?
- What should the royalty rate be? And how does Congress decide what a fair royalty rate is?
- How does one balance the various considerations, such as efficiency and administrative considerations in deciding whether to impose a royalty?
- What would be the economic effects of introducing a system of royalty payments and other rental payments where one did not exist? Would a royalty harm the domestic and international competitiveness of the U.S. hardrock mining industry?
- Finally, what taxes and fees does the hardrock mineral industry pay, and do they have any bearing on the question of whether to impose, or the magnitude of, a royalty?

The first section of this report provides a brief history of the 1872 Mining Law, including the mineral leasing system and royalty payment requirements for the energy minerals. The second section discusses the economic rationale for the payment of royalties on hardrock minerals. It (1) addresses the question of the appropriate economic basis for payment of royalties to the federal government as landowner, (2) compares royalties based on market value with alternative criteria such as royalties based on gross income from depletion and net smelter return, (3) examines the question of an appropriate royalty rate, and (4) discusses some of the likely economic effects of royalty-free provision of public lands, as well as the effects of introducing a royalty. The fifth and final section examines the federal tax provisions, including the three tax subsidies that affect the hardrock minerals industry, as well as claims and patent fees that the industry pays.

Note that the question of payment of royalties from hardrock mineral production on public lands is part of a much broader debate relating to reform of the Mining Law of 1872, which encompasses many other issues — such as whether public domain lands should be patented — which are not addressed in this report.[9]

HISTORY OF THE GENERAL MINING LAW OF 1872

The General Mining Law of 1872 (17 Stat. 91) can be traced to 16th Century English and Spanish property and mineral rights laws and customs, which, as practiced in Colonial America, formed the basis for the granting of the rights to lands including those containing minerals, including a share of the minerals (a royalty).[10] One of the earliest statutes was the "Royal Code of 1783." This Spanish law became the basis for acquiring mineral rights in Spanish colonies in the Americas, and was used to settle claims disputes. Also, the land charters issued by England to the original thirteen colonies provided generally that 1/5th of the minerals on chartered lands were reserved for the crown.[11] Independence for the American colonies meant that issues of land and mineral ownership between the national government, the states, localities, and even private individuals had to be decided. Several ordinances were enacted — the Ordinance of 1785, the Land Act of 1796, and the Preemption Act of 1841 — but these did not address the question of mineral royalties specifically.

The "California gold rush" of 1849 and the silver strike in Nevada of 1860 caused a significant westward population migration for the purposes of prospecting and claiming public domain lands. During this pre-Mining Law period, the issue of land ownership and easy and free access to federal lands was very controversial and fiercely debated. The Congress had approved several temporary leasing or land sales acts for gold, silver, lead, and iron — which were administered by the War Department — but prospectors disregarded them, trespassed onto federal lands, laid their claims anyway, and paid no royalty to the federal government. Also after the Civil War, congressional policy became more openly geared to encouraging westward migration, and the economic development of the West.[12]

These practices and policies became embodied in two statutes that became the precursors to the General Mining Law of 1872: the Lode Law of 1866,[13] which provided free and open access to federal lands with lode deposits for prospecting, claiming, and patenting, and the Placer Act of 1870,[14] which applied the same provisions to placer deposits.[15]

Conveyance and Patenting of Federal Hardrock Mineral Lands

The General Mining Law of 1872, approved by President Ulysses S. Grant on May 10, 1872, was the consolidation of amended versions of both the 1866 Lode Law and the 1870

Placer Act. Under this law, as amended, public lands with hardrock minerals (metals such as gold, silver, copper, lead, etc., and non-metals such as uranium, barite, and fluorspar) are conveyed, rather than leased for a rent and royalty to the federal government as landowner. More specifically, the law permits citizens and businesses to prospect for hardrock minerals on those public lands not withdrawn from mining, and to file a claim, which gives them the right to explore, develop, mine, and sell hardrock minerals from the claim (such as gold, silver, platinum, copper, lead, tin, etc.) without paying a royalty.[16] Further, up until 1994, a claim holder could obtain title — called a patent — to the land and mineral rights upon demonstrating the discovery of economically recoverable minerals. Patenting basically means that the title to the land and mineral rights are conveyed to the claimant — the mineral lands become private property.[17] The claim holder may also claim and patent nonmineral, noncontiguous lands to mill and process the ores. As noted, the General Mining Law was enacted to promote westward migration and the economic development of the West, and the land and mining rushes that followed further accelerated the great Westward settlement and expansion.[18]

Royalties on Leasable Minerals

Originally, the Mining Law applied to all minerals. But over time all but the hardrock or "locatable" minerals were removed from the law's purview.[19] Beginning in 1920, for example, the Mineral Leasing Act of 1920 introduced a leasing system for coal, oil, and gas, and certain other minerals.[20] Each lease was negotiated on a case-by-case basis, and provided for both an annual rental per acre and a fixed royalty of so many cents per ton.[21] Where there is competition, the minerals are alienated through a bonus bidding process. Beginning in 1955, the Mineral Materials Act of 1947 instituted a leasing system for the non-hardrock and non-energy minerals — the common and less valuable minerals such as sand and gravel.

For coal, this system of leasing continued until the early 1970s, when the Arab oil embargo and shortages of petroleum reserves focused attention on the vast coal deposits in the western states. Thus, in 1976 the Federal Coal Leasing Amendments Act (P.L. 94-377) was enacted, which among other changes, replaced the fixed royalty with an ad-valorem royalty of 12.5% for surface mines and 8% for underground mines.[22] This kind of royalty payment system is now also used by the federal government in leasing oil and gas properties, which typically are assessed advalorem royalties ranging from between 12.5% and 16.67%.[23] As will be discussed below, under an ad-valorem royalty — the most prevalent type of royalty system — the royalty payments are based on the value of the mineral. State governments also typically charge ad-valorem royalties for the extraction of locatable minerals, although some tend to use the fixed royalties. Finally, hardrock mining on acquired federal lands (as opposed to public domain lands) is required to pay a royalty of 5%. The royalty rates currently in effect on various types of minerals, both leasable and locatable, from federal lands are shown in **Table 1**.

Thus, despite the evolution of the system for paying royalties on coal, oil, and gas, and certain other minerals, the location system for the hardrock minerals industry — involving no royalty payments or any other type of rental payment — has remained essentially unchanged since 1872.[24]

THE ECONOMICS OF FEDERAL ROYALTIES ON PUBLIC DOMAIN LANDS

What is a Royalty?

Part of the problem in deciding whether to assess a royalty on hardrock minerals from federal lands, and how to structure such a royalty, is confusion over just what a royalty is and what it is not. Economics is very clear on this: A royalty is a factor payment, part of the rent paid, or the return, to land as both a marketable capital asset and input to production. It is a payment made by the renter of the land to the landowner (whether private or public) in a voluntary exchange for the flow of services provided by that land over time. As a factor payment, the royalty is analogous to the wage rate, which is a payment for the services of labor, or the interest rate, which is a payment for the services of capital.

Table 1. Royalty Rates Paid by Mining Companies on Minerals from Federal Lands

Mineral Type or Location	Royalty Rate
Hardrock Minerals on *Acquired* Federal Lands	5.00%
Coal from Underground Mines	8.00%
Coal from Surface Mines	12.00%
Onshore/Offshore Oil and Gas[a]	12.50%
Deepwater Offshore Oil and Gas	16.67%
Geothermal Leases[b]	10-15%
Electricity from Geothermal Leases[c]	1-2.5%

Source: U.S. Department of Interior. Minerals Management Service. *Mineral Revenue Management*. September 2006; and U.S. Department of the Interior. Minerals Management Service. *Mineral Revenues 2000: Report on Receipts from Federal and American Indian Leases*. p. 134.

a. Most federal oil and gas is leased at a 12.5% royalty rate both onshore and offshore; in addition to the above royalties, oil, gas, and coal leases pay rents and a variety of fees.

b. Byproducts from geothermal reservoirs, such as sulfur or zinc, are assessed a royalty of 5%.

c. This is the royalty rate applicable from 2006-2015. The rate applicable after that is between 2% and 5%. The Secretary of the Interior has the discretion to set the actual rate between this range, depending on several criteria, such as revenues.

Mineral producers, as business organizations, require land, as well as labor, capital, energy, and other materials, in order to establish their enterprises and produce goods and services — minerals that provide utility to consumers. And just as mineral producers must pay a wage in exchange for the services of labor, or interest for the service of capital, they generally must pay landowners for the services of land that contains a mineral deposit. The exception to this rule, of course, has been the case of locatable minerals on public (or federal) lands in the United States, on which royalties are not paid.

In the case of mineral lands, rents could be paid in various forms such as a bonus bid, annual rentals, or a royalty, or in various combinations of these depending on the type of mineral, and whether there is a lease or not, and the contractual agreement between a developer of the resources and the landowner. For example, under the Outer Continental Shelf Lands Act of 1953, as amended, the federal government leases lands for oil and gas development in return for a bonus bid, annual rents, and royalties.[25] Lease sales are conducted through a competitive bidding process, and leases are awarded to the highest bidder, who makes an up-front cash payment called a bonus bid in order to secure the lease. Annual rents range from $5 to $9.50 per acre, with lease sizes ranging form 2,500 to nearly 6,000 acres, and royalty rates are, as noted above, either 12.5% or 16.67%.

These mineral rents are an attempt to capture the returns to the land above and beyond the returns paid to labor (wages), capital (interest), entrepreneurship (profits), and other factors, and above any taxes that have to be paid to government. With perfect knowledge and no risks, for example, the rents resulting from mineral lands could be captured by the landowner as up-front payments — as the price of the mineral rights, for example. However, mineral production, like all business, is risky; it is difficult to know in advance of production precisely the quantity and quality of the mineral. There are long lead times between exploration, discovery, and actual production, and it is difficult to project what mineral prices will be upon production and sale. These and other uncertainties make it risky for both the hardrock mineral producer and landowner to predict up front what rents would be earned by mineral lands, and therefore what the mineral producer should pay the landowner. In general, the precise division between royalties or bonus bids and annual rentals depends primarily upon how production risk is shared between landowner and mineral producer. The royalty becomes a way of allowing for mineral land rents to be paid, for the landowner to earn a return on the land, in a way that simultaneously minimizes the risk of either overpayment or underpayment. As a land rental, then, an ad-valorem royalty protects the mineral producer against excessive royalty payments (overestimation of rents) and the government against underestimation of economic rents.

Being a factor payment, then, a royalty is not a tax, which is a compulsory levy on individuals and businesses to finance the cost of government for the common welfare and not a return to a factor of production in exchange for specific services provided. This is an important point, one that might be used, for example, to argue against proposals to impose a royalty based on net profits, which would make the royalty more of an income tax rather than a factor payment.[26] The distinction between a tax and the rental payment also distinguishes between the federal government as a sovereign entity with the power to tax, and as a landowner or proprietor with the right to earn a return on that land for the public.

Economic Implications of Royalty-Free Use of Federal Land

Forgoing royalty and other payments through the location system in effect means that the services of a factor of production are provided free of charge. In this case the services of federal land are provided free to the producer by virtue of the fact that no royalty is ever collected from its use. In economic terms, this is comparable to the free provision of labor or capital equipment. Economic theory suggests that a consequence of providing factor services for free that would otherwise have economic value would be to distort the allocation of resources. This theory holds that if a factor is provided to a firm without charge, too much of that factor would be hired and the output of the firm would exceed the socially desirable amount. Consequently, not enough of a more valuable good is provided.

From another perspective, the federal government is the steward of the public lands which it holds as trustee of American public. The development of public resources should only occur at a fair market value return to the nation, it is argued. An implication of the location system on federal lands is that all economic surplus that may be earned by the industry in the long run is retained by the industry rather than allocated to society at large. These profits are excess or surplus returns to a resource owner above the level required to produce or supply the resource, i.e., above the normal rate of return. In theory, they accrue naturally to the owner of any natural resource but are not necessary to secure the continued supply of the resource; only a normal long-run return is necessary to ensure such a supply. This is an important rationale for the payment of bonus bids and other rental payments. The theory suggests that any pure economic surplus return from a natural resource should accrue to society.

What wWould Be the Structure of an Economically Efficient (and Fair) Royalty?

In general, the economic concept of a royalty as a factor payment implies that the payment should be based on the market value of the producer's output, whether it be hardrock minerals, coal, or oil and gas.[27] For hardrock minerals, however, as a factor payment, the type of royalty that most closely is intended to capture the rents from mineral lands whose future productivity cannot be precisely determined due to risk — variability in output price, unknown quantity and quality of mineral, etc. — is the ad-valorem royalty based on the value or price of the mining firm's output. Under such a royalty, payments are made in installments over the life of the mine, rather than partially up front, and the economic rent is based on the amount of the mineral produced, and the market value or price of the mineral at the mine.[28] Lands producing minerals of higher quality and value, gold for instance, pay a higher royalty amount; those producing lower quality or value minerals, lead for example, pay a lower amount.

It would be inconsistent with the concept of sharing and with the concept of a factor payment in a competitive market for a royalty to be based on other than market value minus the costs of obtaining it. For example, if instead of payments in kind (deer or crops or precious metals) the landowner were to be paid in money, one would expect him to receive the monetary equivalent of the value of the output. Rational landowners would not settle for less than what the deer, crop, or metal is worth because they could always have the deer, crop, or metals taken to market and sold for at least market value. If they wanted less rent, then presumably that would have been negotiated as a smaller share (instead of 1 deer out of 5, it would perhaps be 1 out of 6). Likewise it would not be rational for the renter to pay to the landowner a royalty based on more than market value.

Alternative Bases for Royalty Payments

Although the ad-valorem royalty appears to be economically the most efficient type of royalty, the question becomes to what base specifically should the royalty rate apply? The aggregate value of a commodity being the product of a price (value/unit) and a total output, the questions are: At what point can the value of a mine's output first be determined? Further, since throughout the production stage value is added or increased, at what stage does mining value end, and non-mining value begin?

Production of valuable mineral products from raw hardrock minerals (e.g., metals such as iron, copper, aluminum, gold, lead) generally involves various stages: extraction or quarrying of the raw minerals (e. g., iron ore, bauxite, galena, and nonmetals such as barite, uranium, diatomite, fluorspar, clay, calcium carbonate, etc., that exist in the hard formations of the earth), delivery to a plant or mill in order to prepare for the initial processing, beneficiating or concentrating to derive an ore concentrate (generally the first valuable product), smelting or refining (i.e., additional processing) of the ore concentrate in order to derive the marketable metal or mineral, and the actual marketing or selling of that metal or mineral.[29] These production stages are shown for the typical case or mineral in **Figure 1**.

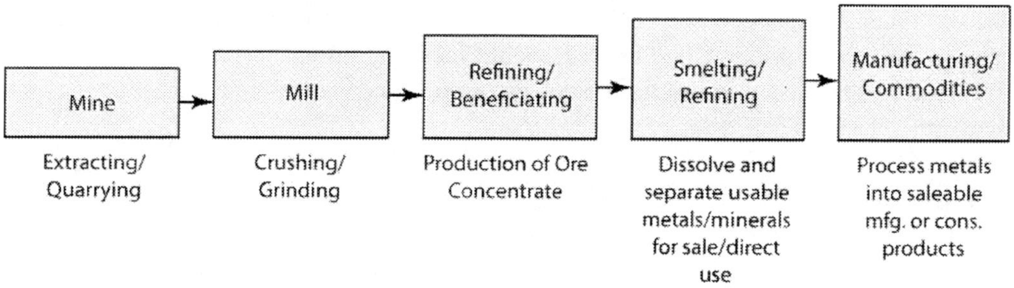

Figure 1. The Various Production Phases for the Typical Hardrock Mineral

Within these production stages, from extraction to the first sale to manufacturers, there may be various points at which there is created a valuable output, depending on the type of mineral, and the structure of the industry — there may be several different types of mineral products with value, and it is not always clear what is mining output and what is non-mining output. Does the mine produce the raw mineral ore, the ore concentrate, or the final metal or mineral sold to wholesalers? Further, even if the output of the mine can be determined, there is the problem of what is the value or price of that output. For example, copper is traded in many forms that relate to different stages of processing: at the mine, copper may be sold as ore, concentrate, or precipitate. This question is further complicated when the firm or industry is vertically integrated. In such cases, there may not actually be a sale throughout the mining-production-processing-marketing (sale) stage, and any price or value would have to be constructed, i.e., estimated or derived, from values further upstream when they are first determined or identified.

Percentage Depletion's "Gross Income" Measure of Mineral Value under the Tax Laws

One measure of the value of mining output is gross income from mining as defined under the income tax laws for purposes of determining the percentage depletion allowance, one of several tax preferences or subsidies available to a mining firm.[30] This royalty base has been suggested many times in the current and past royalty debates and is the base proposed in H.R. 2262. The main reason for suggesting this royalty base is that not only is gross income under the tax depletion rules consistent with the economic concept of an ad-valorem royalty, but the legal and regulatory apparatus has been in place since 1932, the year that percentage depletion first became available to the hardrock mining industry.[31] This would tend to reduce the costs of industry compliance and government administration of the royalty system.

Before discussing this royalty base in more detail, it should be explained that the term "gross income" as used for purposes of percentage depletion is somewhat confusing. There are two reasons for this. While the term is intended to measure only sales revenue from mining operations, income in the context of a business generally refers to profits and not sales. Second, the term "gross" may be misleading because it may suggest a larger percentage depletion deduction, and therefore a larger royalty payment than a "net" royalty. But, in fact, whether a gross royalty base is smaller or bigger than a net royalty base depends on the point in the production stage the price is measured, and the deductions allowed. Because value is added as mining production moves from the extraction or quarrying state to the smeltering stage — the market price, if one existed, would increase throughout these stages — a net price (such as net smelter return) further downstream, would generally be greater than a gross price further upstream.

Definition of Gross Income from Mining

Under Internal Revenue Code (IRC) §613, mining companies are allowed a depletion deduction, at varying percentages, based on the gross income from mining. Under IRC §613(c)(1) and (c)(2), the term gross income from mining means sales revenue from the extraction of the ores or minerals from the ground and any treatment processes necessary or incidental to the mining process, specific treatment processes and also generally the costs of transporting the ores or minerals from the point of extraction to the plant or mill in which the treatment processes are applied. More specifically,

- The term "mining" includes not merely the extraction of the ores or minerals from the ground but also the treatment processes considered as mining described in paragraph (4) (and the treatment processes necessary or incidental thereto), and so much of the transportation of ores or minerals (whether or not common carrier) from the point of extraction from the ground to the plant or mills in which such treatment processes are applied thereto as is not in excess of 50 miles unless the Secretary [of the Treasury] finds that the physical and other requirements are such that the ore or mineral must be transported a greater distance to such plants or mills.

Subparagraph (4) of IRC §613(c) defines, for the different categories of minerals, the treatment processes that when applied by the mine owner or operator are considered as mining (meaning that the costs of such processes are included in the value or price for purposes of percentage depletion):

- "In the case of iron ore, bauxite, ball and sagger clay, rock asphalt, and ores or minerals which are customarily sold in the form of a crude mineral product — sorting, concentrating, sintering, and substantially equivalent processes to bring to shipping grade and form, and loading for shipment [613(c)(4)(C)];"
- "In the case of lead, zinc, copper, gold, silver, uranium, or fluorspar ores, potash, and ores or minerals which are not customarily sold in the form of the crude mineral product — crushing, grinding, and beneficiation by concentration (gravity, flotation,

amalgamation, electrostatic, or magnetic), cyanidation, leaching, crystallization, precipitation (but not including electrolytic deposition, roasting, thermal or electric smelting, or refining), or by substantially equivalent processes or combination of processes used in the separation or extraction of the product or products from the ore or the mineral or minerals from other material from the mine or other natural deposit [IRC §613(c)(4)(D)];"

- In addition, "and so much of the transportation of ores or minerals (whether or not by common carrier) from the point of extraction from the ground to the plants or mills in which such treatment processes are applied thereto as is not in excess of 50 miles unless the Secretary finds that the physical and other requirements are such that the ore or mineral must be transported a greater distance to such plants or mills [IRC §613(c)(2)."

Also under subparagraph (5) of IRC §613(c) the following treatment processes are not considered to be mining processes (and are thus NOT included in gross income):

- "Electrolytic deposition, roasting, calcining, thermal or electric smelting, refining, polishing, fine pulverization, blending with other materials, treatment effecting a chemical change, thermal action, and molding or shaping."

It seems fairly clear from these statutory provisions that gross income from mining is the mining company's revenue from the sale of the first marketable mineral product created as a result of mining and before any non-mining processes (e.g., manufacturing processes such as refining or smeltering) are applied. Generally, for most hardrock minerals extraction produces a raw mineral ore, which of itself has little value and is generally not marketed. For most minerals, after the ore is mined it is sent to a mill or plant to be crushed and ground, and in some cases (lead, for example) is beneficiated to the point that it is saleable as a metal concentrate (e.g., aluminum oxide from bauxite or copper concentrate from copper ore, etc.).[32] At this stage, there is a demand for the mineral ore concentrate as an input into the smelting/refining process for final sale in the wholesale commodities or manufacturing markets (e.g., gold into jewelry, copper for automobiles and piping). Consider gold, for example. Gold refiners are generally the primary market for miners. Raw gold is usually delivered to the refiner where it is purchased from the miner, processed, and refined. At the point of sale to the refiner, funds received are considered gross income to the miner.

After the milling or mining processing stage, the ore concentrate has value due to the value of the mineral to be derived therefrom: copper from copper ore concentrate, aluminum from aluminum oxide, etc. In general, such ore concentrates are either sold to a broker or to a smelter refiner, or they are tolled, which means the ore is not sold, but is rather sent for refining/smelting (and other processing) only — the mining company still owns the ore. If the miner's output is sold in an arm's-length transaction (this would generally occur for certain iron ores sold by independent, i.e., "small" mining companies), then there is a sale of the mineral and that is the actual price used to determine gross income. Thus under IRS regulations, gross income for depletion purposes is defined as "the actual price for which the ore or mineral is sold where the taxpayer sells the ore or mineral as it emerges from the mine before application of any processes other than a mining process or any transportation, or after application of only mining processes, including mining transportation." The price of the ore concentrate would depend on the (wholesale) price of the final metal (including related metals) on commodities markets, and the percentage of the metal in the ore.

If the mining firm is an integrated producer, i.e., it has smelting/refining operations, and also applies non-mining processes to the ore concentrate before the mineral is sold, there may be no sale of either the raw ore, or ore concentrate, and hence no price for the mineral firm's output until the final metal (gold, for example) is sold in the commodities markets or for use in manufacturing. In these cases, percentage depletion is based not on actual gross income but what is called "constructive" gross income, i.e, income derived from or approximated by using representative market prices elsewhere in that particular market. Finally, in cases where there is no representative price from sales of the same ore concentrate (the mining output) in other markets, then gross income from mining is estimated by multiplying gross sales by the ratio of mining to total costs or one of various other methods to estimate the price of the mining output.

As a hypothetical example, assume that the market price of the final cast copper on the commodities exchanges is \$3.60/lb, which is \$7,937/metric tonne of pure copper.[33] Also assume that one metric ton of copper ore concentrate contains 30% pure copper and no other valuable minerals — the remaining 70% is sulfur, iron, and other impurities that have no value.[34] And finally, to simplify further, assume that mining costs represent 75% of the total costs of producing the final pure copper. Then, under the constructive gross income method, gross income per tonne from mining operations — the average sales revenue per unit from the sale of copper concentrate — would be calculated as follows:

$$\$1,786 = (\$7,937/\text{tonne}) \times (0.30) \times (0.75)$$

If the mining company produced 100,000 tons of copper ore concentrate (equivalent to 30,000 tonnes of pure copper, which would eventually be refined into copper ingots or bars) total gross income from mining in this one year would be \$178.6 million. At 8%, the royalty on this would be \$14.24 million.

Net Smelter Return (NSR) Royalty

Another possible royalty base — one which is used in many private, and most state, royalty contracts — is "net-smelter return."[35]

Net smelter return is a common term used in the mining industry to refer essentially to the net average revenue per unit of the smelted mineral product sold by the miner. It is essentially an estimate of the price the miner receives, or equivalently, the price paid by the smelter, for the ore concentrate produced, based on the price for the pure metal (or mineral) received by the refinery or smelter when it is sold in the commodities market. This estimate is based on the (gross) price received by the smelter/refiner for the marketable metal (e.g., the price on the London Exchange or the Chicago Commodities Exchange), less the costs of smelting/refining and other costs (typically treatment charges, penalties for impurities, and transportation charges from the mill to the smelter). According to Cartright (2007),

> The Net Smelter Return is the amount of money which the smelter or refinery pays the mining operator for the mineral product and is usually based on a spot, or current price of the mineral, with deductions for the costs associated with further processing.

In non-metal mines the selling price is usually 'fob mine site' because of the transportation costs involved in delivering the mineral product to the buyer.[36]

In Canadian law, net smelter return is defined in the same way. According to Barten (1993), a net smelter return is

> ... the amount received by the mine or mill owner from the sale of the mineral product to the treatment plant that converts the output of the mill to marketable metal. From the gross proceeds received there may be deductions for costs incurred by the owner after the property leaves the mine property and before sale, such as costs of transportation, insurance, or security, penalties, sampling and assaying, refining and smelting, and marketing. No deductions are made for the operating costs of the mine-mill complex.[37]

Finally, Baldwin (2003) writes:

> A net smelter return will be ... the net value of minerals from the property, where Net Value is Gross Value less certain permitted expenses. Gross value will usually be defined with reference to the spot price of an average weekly or monthly spot price on specified markets for the commodity. The permitted deductions will include all charges by the smelter, such as smelting and refining charge as well as penalties for impurities in the ore, and the cost of transportation to the smelter and insurance. In some cases deductions may also be made for taxes except income taxes and, depending on the nature of the commodity, for marketing costs. If the operator ships to its smelter, charges should not exceed those that would be charged by an arm's length smelter in similar circumstances.[38]

The similarity of these definitions of net smelter return (gross price less nonmining costs) times output to the gross income base under the percentage depletion rules is readily apparent. Each of the above definitions is very specific that smelter and refining charges are to be deducted, just as in the tax definition of gross income.

The following two examples, from contracts or royalty deeds between a landowner (the vendor) and the mining company (the purchaser), illustrate the legal language used in defining a net smelter return:

> As additional consideration for the Vendor entering into this Agreement, the Purchaser acknowledges that the Claims shall be subject to a royalty or charge in the amount of two percent (2%) of net smelter returns payable to the Vendor. For the purpose of this clause "Net Smelter Returns" shall mean the actual proceeds received by the Purchaser from a smelter or other place of sale or treatment in respect of all ore removed by the Purchaser from the Claims as evidenced by its returns or settlement sheets after deducting from the said proceeds all freight or other transportation costs from the Claims, to the smelter or other place of sale or treatment, but without any other deduction whatsoever.[39]

> For the purposes of this Agreement and the payment of the Net Smelter Return, net smelter returns shall mean the actual proceeds received by Warren or any of his respective assignees or successors in interest to the Claims from any mint, smelter,

> refinery or other purchaser from the sale of concentrates, metals (including bullion) or products from the Claims and sold, after deducting from such proceeds the following charges levied by third parties to the extent that they are not deducted by the purchaser in computing payment: a) assay costs and umpire assay costs charged by any mint, smelter, refinery or other purchaser; b) smelting and refining charges, penalties, and the cost of transportation and handling of such concentrates, metals (including bullion) or products from the Claims to any mint, smelter, refinery or other purchaser; and c) related insurance on such concentrates, metals (including bullion) or products from the Claims.[40]

Thus, continuing with the above hypothetical copper example, the net smelter return would be the price at which cast copper could be sold after the costs of smelting and refining the copper ore are subtracted (so as to deduct any non-mining value). Using the same figures as before, the market price of the final cast copper is $3.60/lb, and thus one metric tonne (2,205 lbs) of pure copper would be valued at $7,938. However, one ton of copper ore concentrate contains only 30% pure copper, so that the value of one tonne of that ore concentrate is 30% of that, or $2,381.[41] To calculate smeltering and refining costs assume that these are 10% or $0.36/lb of the price of the cast copper. Then costs would be $238/tonne of copper concentrate (equal to $0.36 x 0.30 x 2,205) the net smelter return would be:

$$\$2,143 = (\$2,381) - (\$0.36) \times (0.30) \times (2,205)$$

Multiplied by the tonnage of ore concentrate shipped to the smelter/refiner, it would equal the actual (or constructive) sales revenue to the mining company of the ore concentrate. If annual sales were 100,000 metric tonnes of copper concentrate, total net smelter return revenues would be $214 million. At 8%, the royalty on this would be $17 million.

Note that in this example, net smelter return is greater than gross income for depletion purposes and would thus yield a greater royalty. This is because final metal prices are high relative to smelting/refining costs. In general, the higher are metals prices relative to costs (in which case mineral producers earn extra or windfall profits) the more net smelter returns diverge from gross income.

Net Profit Royalty

Another royalty base that has been mentioned in the General Mining law debate is net profit. Here the royalty would be a certain percentage of the miner's net after-tax profit or income, which is essentially gross revenue less all costs (operating, capital, and tax costs). For any mining company (indeed for any business) net profit is, of course, less than sales revenue by the costs of doing business, thus resulting in a smaller royalty base from which to calculate royalty payments. Reducing royalty payments to the federal government might be a major reason that it is supported by much of the hardrock mining industry and its supporters. Not only are net profits notoriously variable, difficult to measure, and susceptible to accounting manipulations, but this method could create incentives for mining companies to claim significant deductions from their income such that there would rarely be a profit against which to levy the royalty. Thus depending upon the royalty rate this could lead not only to smaller royalty payments than either a gross income or net smelter return royalty, but perhaps to no royalty at all. Some advocates of the net after-tax profit type of royalty, for example the

1993 bill S. 775, would have measured net profits at the mine mouth, which would generally (depending on the rate) result in even lower royalties. Further, a net profit royalty at the minemouth creates complications in allocating costs between extraction and other mining processing.

Another effect of a net-profit type of royalty on hardrock minerals would be to expose the federal government to the risks of an investor — much like investors in stock are as owners subject to risks — rather than the rewards of the much less risky status of landowner. Such an arrangement would not serve to impose the marketdetermined factor costs (royalties) on the mining operations. In other words, if net profit were to mean net before-tax profits, then the royalty would essentially operate like an income tax. This would not be consistent with mainstream economic theory that a royalty is not a tax, but a factor payment.

The Fixed Royalty

Yet another royalty base, one which is used less frequently, is the royalty based on a constant rate per unit, say $5/ton, of mineral output (i.e., the ore concentrate). This was the type of royalty instituted under the Mineral Leasing Act of 1920, the law which first introduced a federal leasing system, including a royalty, for coal.[42] As a factor payment, this type of royalty makes little economic sense, being equivalent to charging fixed interest payments on capital regardless of the value of the capital loaned. Being assessed on only one variable — output, rather than price and output — it does, however, avoid the usual complexities associated with determining value (price), costs, or even profits.[43]

Comparison with Oil, Gas, Coal, and Geothermal Royalties

For oil, gas, and coal, federal royalties are calculated on the gross proceeds (or gross sales revenues) from the contract sale. For oil and gas this generally occurs close to the wellhead, and so the contract price is used to determine gross proceeds. Typically, for example, oil and gas is sold at a posted price to a pipeline company, and that is the price on which the royalty is based. In situations where oil is not sold in an arm'slength transaction, or is sold without a contract, royalties are based on various types of benchmarks such as comparable sales or even local spot-market oil prices, adjusted for any quality differences and transportation costs. In situations where the oil or gas is transported before it is sold, then transportation charges are generally deducted to determine gross proceeds for royalty purposes.[44]

Royalties on coal extracted from federal lands, at the 8% or 12% royalty rate, are also assessed on the "gross proceeds" from the sale of the coal. This is essentially the price of the coal in a free-market transaction between unaffiliated parties at arm's-length. The price is determined at the point of sale, which for coal is typically at or near the mine-mouth when the coal is weighed and loaded (i.e., at a scale or batch loading facility). Transportation and washing costs, and other costs not deemed to be mining or production costs, are generally deductible. These rules for oil, gas, and coal clarify that value is determined as close to the mine as possible, and that non-mining or extraction costs after a marketable product is created are deductible.[45]

For geothermal resources, the Geothermal Steam Act of 1970 (as amended) instituted a 10%-15% royalty on federal leases based on the value of the steam or any other form of heat or energy produced and sold.[46] However, under amendments made by the Energy Policy Act of 2005 (P.L. 109-58), if the geothermal energy is not sold but is instead used to generate electricity — in most cases it is — the royalty is between 1% and 2.5% and it is based on the gross proceeds from the sale of electricity rather than the value or volume of geothermal energy used in the generation of electricity. In some cases, a deduction against gross proceeds for transmission and wheeling costs is allowed.[47]

What Would Be the Appropriate Royalty Rate?

With regard to a specific royalty rate, economic theory is less clear beyond the implication that a royalty rate determined in the competitive marketplace is generally the most economically efficient rate — the rate that is most likely to maximize social welfare. In the case of privately owned mineral lands, markets generally exist that determine the royalty type and rate for a wide variety of minerals.

In most types of private royalty arrangements in the early 1990s (the latest data readily available), the most common type of royalty was the ad-valorem royalty at rates ranging from 2-8%, with an average rate of 5%.48 In the case of publicly owned lands, laws determine the return on the resources, although competitive market rates may be a determining factor in establishing such rates. Most states with mineral resources imposed ad-valorem royalties at rates ranging from 2-10%.[49] For leasable energy minerals on federal lands, the statutory royalty rates range from 5%-16.67%. For oil and gas, the royalty rate is either a 1/8 (12.5%) or 1/6 (16.67%) share of the value of the mineral, depending upon whether the oil or gas is shallow (1/6 share because costs are lower) or deep (a 1/8 share because costs are higher). On some leases, the rate could be higher than 1/6. Also, the royalty could be paid "in-kind" (either a 1/8 or 1/6 share of the output rather than of the price). For coal, the royalty rate is either 12% (surface mines) or 8% (underground mines). Note that the 8% advalorem rate proposed in H.R. 2262 is the same as the royalty rate on underground coal mines. Even for hardrock minerals on acquired lands (as opposed to public domain lands, which are governed by the 1872 Mining Law), the Congress has established an ad-valorem royalty rate of 5%.[50] Finally, in international lease transactions, mineral royalties are predominantly of the ad-valorem type with rates ranging typically from 2%-12%, depending on the country, and the mineral type.[51]

External Benefits

There may be conditions under which the government might not want to use the royalty rate established by the market. If there are sufficient positive economic and social effects from the use of public lands for mining, there might be an economic rationale for charging the mineral producer a lower than market rate.[52] Such externality arguments — the stimulation of an infant industry and the development of mineral resources, and the settling of the American West — were apparently a principal rationale for the location system involving no royalties and other rent payments established in 1872. But these arguments may not be valid in today's economy. The hardrock minerals industry is not an infant industry, and there is no longer a need for settling the American West.

Arguments are made that the economic stimulus to the industry resulting from royalty-free treatment produces benefits; it is argued that imposition of a federal royalty on hardrock mining would damage most of the American hardrock industry, which, it is argued, suffers from fierce, heavily subsidized foreign competition. Some argue that there is a national security aspect to subsidizing certain hardrock minerals, although there is no presumption that royalty-free treatment is the appropriate way to address these concerns. It is also pointed out that damage to the industry would harm those western communities that are economically dependent upon it.

These arguments are contested on various grounds. Subsidies for hardrock mining clearly produce economic benefits to the industry receiving the subsidy and to the communities dependent upon that industry. But these benefits are not external benefits that the markets are not capable of taking into account; they are not "market failures," which may justify either a tax (in the case of external costs) or subsidies in the case of external benefits. Rather, the benefits are *private* benefits that are taken into account by the normal working of the market system. Further, it can be argued that the private benefits to the hardrock mining industry occur at the expense of private damages to other American industries; the private benefits to miningdependent regions of the United States may occur at the expense of private costs to other regions of the United States. When resources are scarce, a condition that affects all economic systems and all countries alike, the use of resources in one sector means the loss of the resource in another sector — the subsidy to hardrock mining is not costless. The loss of economic efficiency implies that the private benefits to the hardrock mining industry are smaller than the private costs to other sectors. This suggests that removal of the subsidies would produce a net welfare gain to citizens of the United States.

As to the national security argument, there are minerals that are often referred to as "strategic minerals" — cobalt, columbium, chromium, manganese, platinum, and others.[53] Excessive dependence upon foreign producers of these and other economically vital minerals might pose an economic disruption or national security threat to the United States, although the nature of the relationship between imports, domestic production, and national security is not a simple one. These arguments are similar to those made about the U.S. dependence upon imported oil, and often used to justify subsidies for the domestic oil industry. Some argue that the threat of a domestic supply disruption might be better met with a stockpiling of that mineral.[54] It seems clear also that not all hardrock minerals are of U.S. national security concern. In any event, if a financial subsidy for certain strategic hardrock minerals industries was appropriate, a tax subsidy would probably be more efficient than total forgiveness of royalty payments. (The current tax subsidies for the hardrock mineral industry are discussed in the section "Federal Tax Treatment of the Hardrock Mining Industry," below).

External Costs

There is another side to the externality concept: external costs. It appears that some external effects associated with hardrock mineral production — economic effects not taken into account in the normal business decisionmaking — are net "external costs" rather than net external benefits. The external costs from hardrock mineral production are the degradation of the environment and the damage to the utility of the land for other potential uses. Without land reclamation, there would presumably be damage to the environment, which is an external cost to society. This is a cost that the typical profit-maximizing hardrock mineral producer does not usually account for in the normal business decisionmaking process. Consequently, since the accountable business costs are lower than the real costs (including external costs) prices are lower than they should be, causing output to be higher than it should be. Under this reasoning, if the presence of external benefits is dismissed and the presence of external costs is considered significant, the appropriate federal policy would be a tax rather than to impose no royalty on the use of the land. This is the economic rationale for the abandoned mineland reclamation fee under SMCRA, the Surface Mining Control and Reclamation Act.[55]

Absence of Competition

Two other conditions might mitigate against the use of the market-determined royalty rate: First if the market for the leasing of mineral lands between landowners and mining companies is not competitive, and second, if the mining industry itself is not competitive. In the first case, if private landowners exercise monopoly power or some other type of market power, then it would be possible to extract from the mining companies higher than competitive royalty rates — monopoly rates. In the second case, if the mining industry is concentrated and exercises market power — control over prices, for example — it might lead to underpayment of royalties — the market-determined royalty rates might actually be too low.

These are complex issues which are difficult to analyze. Much of the land from which hardrock minerals are produced is owned by the federal government, which might be in a position to take advantage of its power to obtain royalty payments above the competitively determined rates. Such payments, combined with the bonus bids, would appropriate for society the economic rents — the abnormal or excess profits (i.e., windfalls) — from the production of scarce natural resources. In reality, of course, the federal government policy is just the reverse: It charges a royalty rate of zero, which is definitely below the market-determined rate. On the other hand, much land is also privately owned or owned by state or local governments or Indian tribes, which suggests that the market-determined royalty rates might be a reliable indicator of competitive rates.

As to the hardrock mineral industry structure, available data suggest that the industry, both in the United States and throughout the world, is somewhat concentrated. Hardrock mining is a highly capital intensive business and large enterprises can operate more efficiently. In the United States, many of the mining firms are large multinational mining conglomerates — five of the top ten claim holders (who hold more than 16% of all claims on western public lands) are foreign corporations. Internationally, the hardrock mining industry is dominated by about 40 companies and the top four account for 35% of total market capitalization.[56] This would suggest that the market-determined royalty rate might actually be too low. On balance, however, it might be the case that the above two opposing forces tend to offset each other and the government would want to use the more appropriate competitive rate, which would be more consistent with economic efficiency, and social welfare, than a zero rate.

Should Royalties Rates Differ by Mineral Type?

Economic theory is also unclear whether the same royalty rate should apply to all hardrock minerals or whether the rate should vary by type of mineral in some logical way — according to some acceptable rational criteria. On the one hand, a variable rate is unnecessary because the base is ad-valorem, so the royalty amounts, in absolute dollars, automatically vary in proportion to mineral value: more valuable minerals pay greater absolute royalties per unit. **Table 2** gives an illustration for metals. Note that the royalty in column (4) is based on the gross value of the metal in the commodities markets and does not allow deductions for smeltering costs and other costs — in other words, the figures do not represent either net smelter returns or gross income from mining, both of which would be less by the amount of such costs. A constant royalty rate also tends to minimize both economic distortions — intra-industry resource allocations among mineral types — as well as administrative costs. On the other hand, private royalty agreements, which are negotiated on a caseby- case basis, have widely varying royalty rates. These can range from 1-10%, and differ by mineral types.

Under one possible structure that has been suggested, the royalty rate itself would be a function of — it would vary positively with — the mineral's price. Column (6) of **Table 2** shows one hypothetical schedule of such rates beginning at 8% for platinum, the most valuable metal, and declining in steps of 1 percentage point to 1% for aluminum. With such a schedule the absolute amount of royalty payments declines even more rapidly as the value of a mineral declines.

Another possibility would be to vary the royalty rates in direct proportion to the percentage depletion rates applicable to gross income from mining as shown in **Table 3**. Such a royalty rate structure would produce a different pattern of absolute royalty payments than either a constant royalty rate or a royalty rate in direct proportion to the minerals value. For example, lead and zinc, which qualify for a 22% depletion rate, would have a higher royalty burden than gold or silver, which are significantly more valuable than either lead or zinc but which qualify for a 15% depletion rate.

Effects on the Mining Industry

Introducing a royalty payment system for locatable minerals could have an adverse economic and financial effect on the producers of these minerals, particularly on marginal producers. Both average and marginal costs would increase, and profits and employment in the hardrock mineral industry would decrease. Hardrock mineral production would be lower as marginal mines closed. This would also adversely affect those parts of the United States that rely on these industries. The payment of royalties combined with higher fees could also alter the competitive balance among mineral producers both domestically and internationally: The competitive disadvantage that royalty-paying mineral producers currently have would be lessened. The hardrock mineral industry generally would be less competitive in relation to other domestic mineral producers and with international mineral producers. Hardrock mineral prices would probably not increase very much because for most minerals, prices are determined in the international market — they cannot be shifted. This would further add somewhat to financial pressures for the marginally profitable mining company.

On the other hand, the hardrock mining industry has been booming over the last four years. One indicator of this is the growth in metals and minerals prices, which have increased overall and have contributed to industry profitability. **Table 2** shows this increase between 2002 and October 2007 for selected metals. According to the Department of Commerce, relative to its size in the economy, the mining industry group generally was the largest contributor to growth in the Gross Domestic Product price index in 2006, accounting for

6.6% of price growth, more than three times its share of current-dollar GDP.[57] According to PricewaterhouseCoopers, 2006 was another spectacular year for the global mining industry.[58] It also predicted that 2007 should bring with it both record financial results and further consolidation within the industry. Finally, rates of return on capital for the U.S. mining industry have also been higher as compared with many other industries.[59]

Table 2. Illustration of Ad-Valorem Royalty Calculations based on Recent Minerals Prices

Mineral Type	Price in 2002 ($/lb)	Price on 10-17-2007 ($/lb)	Constant Royalty at 8% ($/lb)	Constant Royalty at 4% ($/lb)	Variable Royalty Rate (% and $/lb)
(1)	(2)	(3)	(4)	(5)	(6)
Platinum	$6,516.00	$17,400.00	$1,392.00	$696.00	8% = $1,392.00
Gold	$3,732.00	$9,000.00	$720.00	$360.00	7% = $630.00
Silver	$55.00	$174.00	$13.92	$6.96	6% = $10.00
Tin	$1.95	$9.70	$0.78	$0.39	5% = $0.49
Copper	$0.75	$3.60	$0.29	$0.14	4% = $0.14
Lead	$0.44	$2.00	$0.16	$0.08	3% = $0.06
Zinc	$0.38	$1.50	$0.12	$0.06	2% = $0.03
Aluminum	$0.64	$1.10	$0.09	$0.04	1% = $0.01

Sources: U.S. Department of the Interior. U.S. Geological Survey. *Mineral Commodity Summaries: 2007;* and the Wall Street Journal, October 17, 2007. p. C8.

Other Economic Implications of Imposing a Royalty

These negative industry and regional effects would be a natural and unavoidable part of the transition from a policy of resource subsidization to a policy of resource neutrality. As the above discussion suggests, however, the policy of introducing a royalty payment system for locatable minerals would be expected to have positive economic effects in other sectors (the non-mining industry) and in the remaining parts of the country. Less land would be mined and fewer U.S. hardrock minerals would be produced and more imported than under a no-royalty system, but more of other, higher-valued commodities would be produced. This would likely contribute to a more economically efficient use of the nation's resources, which would tend to promote the welfare of society — a greater good for a greater number of people.

Thus, the current policy of free access (i.e., royalty forgiveness) involves a trade-off: Policymakers choose the benefits to the hardrock mineral industry above the benefits that would accrue to another domestic economic sector. A policy of royalty (or other rental) payments would be more neutral; it would let the markets be the arbiter of the relative size of the various economic sectors. And even if the health and competitiveness of the domestic hardrock mineral industry is of sufficient concern to warrant federal subsidies, there may be more economically efficient ways of providing these subsidies than the total forgiveness of a factor payment.

Revenue Effects

The payment of a royalty to the federal government for the services from the mineral land would increase federal revenues, although revenue generation is not the primary rationale for the imposition of a royalty. The Congressional Budget Office has estimated the budgetary effects, revenue and spending, of H.R. 2262. The estimates are replicated in **Table 3,** which shows both revenue effects of the 8% and 4% royalty and from the proposed higher fees.[60]

As noted, H.R. 2262 would establish, beginning after its enactment, a 4% royalty on existing claims, and an 8% royalty on new claims. The 4% royalty on existing claims would generate revenues early, although no royalties would be paid in 2008 due to a provision in H.R. 2262 that would defer royalties during the first 12 months after enactment. Most of the revenues estimated under H.R. 2262 would come from existing claims. The 8% royalty on new claims is unlikely to generate significant revenues during the forecast period (2008-2017) because after a claim is staked, it typically takes at least 10 years for a hardrock mine to begin production. H.R. 2262 would also reauthorize and raise both the claim's location fee and the annual maintenance or holding fee, from $30 to $50 per claim, and from $125 to $150 per claim respectively. The projected revenue effects from the proposed higher fees are shown in **Table 3**.

The revenue estimates are net of federal income taxes — there would be some decline in income tax revenues due to the deductibility of royalty payments and fees against business taxes. Also, note that under federal budget rules fees collected by federal agencies, such as the Department of the Interior, are considered as offsetting budget receipts, or negative spending. Such receipts reduce the agencies' outlays.

Table 3. Revenue Effects of a Royalty on Hardrock Minerals under H.R. 2262 (dollars in millions)

Type	2008	2009	2010	2011	2012	2013	2014	2015	2016	2017	2008-2017
Royalties	0	70	30	30	30	30	30	30	30	30	310
Fees	10	55	51	47	43	40	37	35	33	31	382
Total	10	125	81	77	73	70	67	65	63	61	692

Source: Congressional Budget Office. H.R. 2262: *Hardrock Mining and Reclamation Act of 2007.* CBO Cost Estimate. October 29, 2007.

FEDERAL TAX TREATMENT OF THE HARDROCK MINING INDUSTRY

The U.S. hardrock minerals industry is subject to the same income tax laws which apply to all other for-profit businesses, paying federal income, payroll, and other federal taxes. The federal corporate income tax applies a 35% rate to most corporate taxable income, defined generally as gross revenue (sales) minus interest, wages, the cost of purchased inputs, and an allowance for depreciation. Reduced tax rates, either 15%, 25%, or 34%, apply to "smaller" corporations, those earning less than $10 million of income.[61] In addition, corporate stockholders pay individual income taxes on their corporate-source capital gains and dividends.

As business entities, hardrock mining companies benefit from many different types of untargeted tax incentives, provisions that are generally available to nonmining businesses, as well as some tax provisions specifically targeted toward mining. Two untargeted or general tax benefits are the §199 manufacturing deduction and accelerated depreciation. Under IRC §199, "manufacturing" businesses are allowed a deduction, as a business expense, for a specified percentage of the qualified production activity's income subject to a limit of 50% of the wages paid that are allocable to the domestic production during the taxable year. The deduction was 3% of income for 2006, is currently 6%, and is scheduled to increase to 9% when fully phased in by 2010.[62] For the domestic mining industry (which includes the oil and gas industry), the deduction applies to any primary product "manufactured, produced, or *extracted* in whole or in significant part in the United States." Note that extraction is considered to be manufacturing for purposes of this deduction, which means that domestic firms in the business of extracting hardrock minerals from underground reservoirs or deposits qualify for the deduction.[63] Effectively, the deduction is equivalent to a reduced marginal tax rate. For example, at the top marginal corporate tax rate of 35%, which typically applies to large corporations such as hardrock mining companies, the current deduction of 6% is equivalent to a marginal corporate income tax rate of 32.9% (35% x 0.94) rather than 35%.[64]

Another untargeted or general tax provision that is important to the hardrock mineral companies is accelerated depreciation — the rules that determine how much a business may deduct for the depreciation of its capital equipment and structures. This deduction is intended to provide a business reasonable allowance for the exhaustion and wear and tear (i.e., depreciation) of business property. It is analogous to the cost depletion allowance for investment in a mineral deposit (as discussed below). The present depreciation treatment is known as the Modified Accelerated Cost Recovery System (IRC §168). Generally, under this system, assets are depreciated for regular tax purposes over a specified recovery period, and by applying one of the accelerated depreciation methods prescribed for each class of property, including the 200% declining balance method (the double declining balance method) or the 150% declining balance method or formula.

The annual depreciation deductions are determined by applying this formula to the assigned recovery period. The applicable recovery periods are prescribed for each property class. Mining equipment, assets used in the mining, quarrying, milling, and beneficiating of metallic and nonmetallic minerals, are assigned a seven-year recovery period. This system also allows producers to use a depreciation method (as opposed to the recovery period) that maximizes the value of the deduction over time. This is more advantageous than the theoretical treatment, which would call for a depreciation schedule based on the actual economic depreciation — a longer recovery period and a less accelerated formula, which increases the tax value of the deduction (or equivalently, reduces the effective marginal tax rate on the income generated by the equipment). Given that hardrock mineral producers are highly capital intensive and that the equipment used (stripping shovels, drills, bulldozers, drag lines, and haulers) can cost tens of millions of dollars, the accelerated depreciation provision can be relatively important.

Three Special Tax Preferences or Subsidies

In addition, at the federal level there are three special, or targeted, tax preferences available to the mining industry generally, that are also available to hardrock mining firms.

Expensing of Mine Exploration and Development Costs

One important special tax break available to the hardrock mineral industry is expensing of exploration and development expenditures, i.e., the deduction in the year incurred of costs that are essentially capital in nature (IRC §616 and §617).

Before a mine reaches the producing stage, mining companies incur geological and geophysical (or surveying, and reconnaissance investigation) costs, costs for ascertaining the existence and location of a mine, and costs for determining the quantity and quality of the deposit. These include the costs of staking the claim, removal of property line obstructions, limited removal of overburden (the removal of large amounts of overburden would indicate that a deposit may have already been found and the mine may be in a different stage), and limited sluicing. Exploration expenditures include the costs of surveying, test pitting, trenching, drilling, driving of exploration tunnels and adits, and similar types of work. The costs of exploratory drilling to ascertain the extent of commercially marketable ores are treated by the tax law as a mineral exploration cost. Expenditures for exploratory drilling from within a producing mine to ascertain the existence of what appears (on the basis of all the facts and circumstances known at the time of the expenditure) to be a different ore deposit are also considered exploration expenditures. These types of costs are all deductible rather than capitalized.

Expenditures made after the existence of ores or minerals in commercially marketable quantities has been disclosed are considered mine development expenditures, and these also qualify for expensing treatment. These are costs associated with bringing the mine to production. For example, the costs of constructing or sinking a mine shaft, or the costs of driving tunnels and galleries, are development costs, which may be expensed. Activities associated with development are building roads, clearing the land, and other activities to prepare a site for the production stage.

There are several limitations to expensing of mine exploration and development costs. First, expensing applies only to domestic mines; exploration and development cost of foreign properties must be either depleted or amortized over a 10-year period using straight-line amortization. Second, corporate mining companies may only expense 70% of the exploration and development; the remaining 30% must be amortized over five years, i.e., deducted evenly over 60 months (IRC §291 (b)). Third, both exploration and development costs must be recaptured (i.e, taken back) once the mine is opened and production commences (for exploration expenses) or when the mine is sold (for development expenses). Recapture is accomplished by either including the amount expensed as ordinary income or by subtracting it from the percentage depletion allowance. This makes it especially important for mining companies to determine when exploration ends and development begins.

The Percentage Depletion Allowance

Once a mine reaches the producing stage, a producer may deduct a certain percentage of the gross income — as specified by law — as an allowance for depletion. The percentage depletion allowance is a tax deduction against the federal income tax available to a mining firm for the costs of depleting a mineral reserve, which is part of the total production costs deductible to derive net taxable income. The depletion allowance or deduction is intended to give the mineral producer a deduction for the depletion of the investment in the mine, to account for the loss (depreciation) of capital, to reflect the physical and economic depletion of the mineral reserve as the mineral is being extracted (i.e., the wasting of the asset). It is analogous to the allowance for depreciation of equipment and structures and other assets that is available to all businesses. In theory the depletion allowance should be based on the actual investment cost of the mine, and the annual deduction should be based on the actual output of the mine (its actual rate of depletion). This is called cost depletion.

Table 4. Percentage Depletion Rates for Minerals

22% Depletion Rate		
Any....	If from deposits in United States:	The following metal ores:
Sulfur, and uranium	Anorthosite, asbestos, bauxite, block steatite, talc, celestite, chromite, clay, corundum, fluorspar, graphite, ilmenite, kyanite, laterite, mica, nephilite syenite (to the extent that alumina and aluminum compounds are extracted therefrom), olivine, quartz crystals (radio grade), rutile, and zircon	Antimony, beryllium, bismuth, cadmium, cobalt, columbium, lead, lithium, manganese, molybdenum, nickel, platinum, platinum group metals, tantalum, thorium, tin, titanium, tungsten, vanadium, and zinc
15% Depletion Rate		
Oil and gas	Gold, silver, copper, iron ore, and oil shale from deposits located within the United States	
14% Depletion Rate		
All metal mines not in the United States, and all metal mines in the United States, but not qualifying for the 22% rate	Rock asphalt, and vermiculite, granite, limestone, dimension or ornamental stone, and all other minerals not listed in this table	If not mined in the United States, bauxite, fluorspar, most types of ball clay, bentonite, china clay, sagger clay, and clay used or sold for use for purposes dependent on its refractory properties
10% Depletion Rate		
Asbestos not mined in the United States	Brucite, coal, lignite, perlite, sodium chloride, and wollastonite.	
7.5% Depletion Rate		
Clay and shale used or sold for use in the manu-facture of sewer pipe or brick	Clay, shale, and slate used or sold for use as sintered or burned lightweight aggregates	
5% Depletion Rate		
Clay used, or sold for use, in the manufacture of drainage and roofing tile, flower pots, and kindred products	Gravel, peat, pumice, sand, scoria, shale and stone (except oil shale and stone described elsewhere in this table)	Bromine, calcium chloride, and magnesium chloride from brine wells

Source: Internal Revenue Code Section 613(b).

Unlike the annual depreciation deduction, however, which is calculated based on a taxpayer's actual investment cost of the property (equipment and structures), a recovery period corresponding to that property, and an annual rate of deduction over that recovery period, the percentage depletion deduction for a mine is calculated as a percentage of "gross income" (annual sales revenue) irrespective of the actual investment in the mineral reserve (the mine). That is why percentage depletion is a tax subsidy, preference, or tax expenditure. Any depletion deduction in excess of cost depletion is considered by economists to be a tax subsidy or incentive because the present value of the total amount deducted typically exceeds the initial investment. In other words, the total amounts that can be recovered through percentage depletion generally exceed the total investment cost of the property. This is true even with the net-income limitation, which limits the deduction to 50% of the taxable income from the mine (IRC §613 (a)). Using cost depletion, the total cost recovered cannot exceed the actual investment in the property.

Table 4 shows percentage depletion rates for most minerals, including hardrock minerals. As this table shows, the allowance is available to virtually all mineral producers except integrated oil and gas producers, although the highest rates available are for the hardrock minerals. Percentage depletion rates range from 5% for stone and clay to 22% for uranium and beryllium. Most of the hardrock minerals have a depletion rate of 22%, and several have either a 15% or 14% rate.

The percentage depletion allowance is also subject to several limitations. First the amount of deduction is limited to 50% of the net income (essentially, the net profit) from the property. Second, for iron ore (and coal) mined by corporations (as opposed to noncorporate firms) the allowance must be reduced by 20% of the excess of percentage depletion over cost depletion (IRC §291). This limitation was introduced by the Tax Equity and Fiscal Responsibility Act of 1982 (P.L. 97-248). In effect this limitation reduces the 22% depletion rate to an effective depletion rate of between 17.6% and 22%, depending upon the amount of cost depletion.[65]

The Interaction between the Percentage Depletion Allowance and the Royalty

As noted above, the imposition of a royalty, where none existed before, would increase a mining firm's operating costs and reduce profitability. Further, due to the deductibility of the royalty (as an operating cost of doing business) against the federal income tax, the deduction reduces tax liability and therefore lowers the tax costs — the after-tax cost of the royalty is effectively reduced.

However, there is an important interaction between a royalty and the percentage depletion allowance that limits that effect, i.e., that *increases* the after-tax cost of the royalty somewhat. Under IRC §613, royalty payments (and other rental payments) are deductible from gross income to determine the percentage depletion allowance. This means that if a royalty were to be imposed on hardrock mining from federal lands, it would have the effect of simultaneously reducing the amount of the percentage depletion deduction. Thus, while the imposition of a royalty would increase deductible expenses and reduce the mining company's tax liability, the reduced percentage depletion allowance would reduce tax deductions and would raise a mining firm's tax liability. The net effect on tax liability of these two simultaneous considerations are shown in the following equation, which shows the total increase, after tax burden, for every dollar of royalty paid:

$$= r\,(1\text{-}t) + r\,t\,d \quad (1)$$

$$= r - r\,t + r\,t\,d \quad (2)$$

$$= r - rt\,(1\text{-}\,d) \quad (3)$$

where r = the royalty rate,
t = the marginal corporate (or business) tax rate,
d = the applicable percentage depletion allowance rate.

The first term in (1) shows how the burden of a royalty is reduced by its tax deductibility; and the second term shows how the burden of a royalty is increased by its subtraction from gross income. For example, an 8% royalty without the gross income offset is effectively a 6% royalty at a 25% marginal tax rate; the same royalty however is effectively a 6.44% royalty with a gross income offset if the percentage depletion allowance is 22%. Similarly, a 4% royalty without the gross income offset is effectively a 3% royalty at a 25% marginal tax rate; the same royalty, however, is effectively a 3.15% royalty with a gross income offset if the percentage depletion allowance is 15% (say, for gold).

Equations (2) and (3) are the same as equation (1). Expressed as equation (3), the effective royalty is shown as a gross royalty, the r term, and the offset term - rt (d-1). So in the previous example when the gross royalty is 4%, the revenue offset is a negative 0.85% [0.04 x 0.25 (1 - 0.15) = - 0.85]. This shows that under current tax laws, and given the current percentage depletion rates, the costs of gross income deductibility of the royalty could, in certain cases, offset the benefits from royalty tax deductibility — which, for these cases, is effectively equivalent to not allowing tax deductibility.[66]

Deduction for Mine Closing and Reclamation Costs

The Deficit Reduction Act of 1984 (P.L. 98-369) included a provision that gives preferential tax treatment to the mining industry in the area of mine closing and land reclamation. This provision (IRC §468) allows mining companies to deduct the costs of mine closing and land reclamation in advance of the actual closing and reclamations, i.e., before the occurrence of the activity giving rise to the expenses. This provision is contrary to the general tax rule under both the cash method of accounting and the accrual method of accounting, which state that expenses to be incurred in the future cannot be deducted currently. Under the cash method of accounting, expenses can only be deducted when the payment is actually made. Under the accrual method of accounting, future expenses may be deducted in advance of actual payment but not in advance of the economic activity that gives rise to the expenses — they must be deducted in the year in which the economic activity takes place.

Revenue Losses from Special Tax Preferences

The Joint Tax Committee estimates the revenue losses from the above three special tax preferences or subsidies (formally called "tax expenditures"). The latest five-year estimates (from FY2007-2011) are as follows: $0.5 billion for expensing of mine exploration and development costs, $1.1 billion for the excess of percentage over cost depletion, and $0.2

billion for mine reclamation reserves deduction.[67] Thus the total revenue loss from these three provisions over five years is $1.8 billion, an average of $360 million/year. This is the amount by which the industry lowered its tax burden, or conversely, the loss in tax revenues to the federal government. The bulk of these preferences is claimed by corporations — only a small share is claimed by proprietorships and partnerships.

Alternative Minimum Tax

There is another provision of the current IRC that attempts to limit the benefits from the special tax breaks available to the mining industry and other industries: the alternative minimum tax. This alternative minimum tax is an additional tax on items of tax preference, which are a list of ten tax breaks or subsidies that have been identified as potentially able to reduce a business's or individual's tax liability to little or nothing. The intent is to assure that all taxpayers — individuals and businesses — pay some tax, which is not assured if taxpayers are able to claim a lot of special tax breaks.

In effect, the alternative minimum tax imposes an additional tax above and beyond the normal or regular income tax on the total amount of these ten items of tax preference. This is important for the mining industry because two of the ten preference items are tax breaks that are important to the mining industry: the excess of percentage depletion over cost depletion, and the excess of the exploration and development cost deduction over a hypothetical deduction that would have been allowed if the costs had been amortized ratably (deducted evenly) over a ten year period. A third item of tax preference that the mining firm would have to include in its list of tax preference items is the excess of accelerated depreciation over straight line depreciation.

To the extent to which mining companies uses these and other items of tax preference, this could trigger the minimum tax, and thereby reduce the value of the special tax provisions.

Effective Tax Rates

These special tax preferences or subsidies, combined with accelerated depreciation (a significant tax benefit for highly capital intensive business such as hardrock mining) have historically resulted in relatively low effective average and marginal tax rates. Thus, firms that mine hardrock minerals on public domain lands pay no royalty, and benefit from fairly significant tax subsidies. For example, CBO recently published a study of effective tax rates on different types of assets, and those used in mining (mining and oil field machinery, and mining structures such as mine shafts) generally ranked fairly low, relative to assets used in other industries.[68]

Fees Paid by the Hardrock Mining Industry

Finally, mining companies pay a variety of claims fees (location fees, Bureau of Land Management processing fees, annual maintenance fees). These are charges for specific types of administrative services provided by the BLM. The claims location fee is presently $30 per claim, which is about 20 acres. This fee is charged for first-time locators to locate and record a claim. The annual maintenance (or holding) fee, necessary to hold a claim on public land, is presently $125 per claim. This fee applies to claim holders with 11 or more claims (those with 10 or fewer claims are exempt). Both the location fee and the annual maintenance fee are adjusted for inflation. The BLM also assesses a processing fee of $15. These fees help the BLM cover its costs of administering the mining law and regulations. They cover costs for such actions as processing and recording applications, certain exploration permits, name changes, and corporate mergers, as well as lease consolidations, transfers, and reinstatements.

APPENDIX: DESCRIPTION OF H.R. 2262

On November 1, 2007, the United States House of Representatives approved a substitute version of H.R. 2262 (with amendments), the Hardrock Mining and Reclamation Act of 2007. This bill would:

- establish a new regulatory framework for administering permits to develop hardrock minerals;
- permanently end the sale or "patenting" of public lands for mining, which has been under moratorium since the early 1990s;
- establish royalty payments for hardrock mining operations on federal lands and use 75% of the proceeds for the cleanup of abandoned sites;
- impose an 8% gross income royalty future hardrock mining operations, and an 4% royalty rate on current mining operations;
- require miners to seek additional permits to explore for and develop mineral resources and meet certain standards related to reclamation of mined lands;
- create an abandoned mine reclamation program similar to the one currently in place for coal sites;
- make the income from hardrock mining fees and royalties available, subject to appropriation, to support reclamation programs and to provide assistance to certain state, local, and tribal governments; under an amendment to the Committee-approved bill, 50% of the cleanup funding would be directed to the state in which the royalties were generated;
- direct the Secretary of the Interior to prioritize reclamation projects that protect public health and safety, particularly from water pollution, and for projects that restore wildlife habitat; under an amendment to the Committee bill, watershed areas would be designated as eligible reclamation projects and among the top priorities for receiving cleanup funds;
- establish environmental standards specific to the hardrock mining industry. Environmental standards in the substitute version of H.R. 2262 would be less prescriptive and more performance-oriented than those in the original bill;[69]
- establish a new definition of "undue degradation" of public lands and resources and give the Secretary of the Interior broader authority to block mining operations that would cause such harm;

- allow Native American tribes to petition the federal government to withdraw federal lands with cultural and religious values from mining activities (similar language has been in place for states and localities);
- set guidelines for administrative and judicial review of actions authorizing mining on public lands and allow citizen lawsuits challenging federal decisions.

End Notes

1 19 Stat. 91, 30 U.S.C. §§ 21-54.

2 The term "public domain lands" refers to the original endowment of territory (also called the general territory) comprising the United States. Acquired lands are those lands that have been purchased by, or given to, the federal government, including lands obtained through condemnation. The sum of public domain lands and acquired lands comprises federal lands.

3 Executive Office of the President. *A Vision for Change for America.* House Document 103-49, 103rd Cong, 1st Sess. February 17, 1993. p. 78.

4 In 1993, the Senate was considering S. 775, which proposed a 2% net profits royalty, and the House was considering H.R. 322 (Representative Rahall), which proposed an 8% net smelter return royalty.

5 The two major bill were S. 775 and H.R. 322. See CRS Report 93-632, *Reforming the General Mining Law of 1872: A Comparison of S. 775, and H.R. 322*, by Marc Humphries.

6 A royalty based on "net smelter return" (either mineral price or profits) attempts to determine or assess the royalty payments — the amounts paid to the landowner — on the value of the mineral after it is separated or smelted from its ores. The term is somewhat misleading and will be discussed in detail in the text.

7 The original bill provided for only an 8% royalty; the bill approved by the full committee provides for the two-tiered royalty: 8% on future mining operations and 4% to existing mining operations.

8 "Locatable minerals" refers to the hardrock minerals subject to the Mining Law of 1872 i.e., "those for which the rights are initiated by the location, recordation, and maintenance of a mining claim." "Leasable minerals" refers to coal, oil/gas, and other energy, chemical, or soft minerals.

9 For example, under the patenting system, fee title to mineral lands was conveyed to the mineral companies or claimants at a price of either $2.50/acre or $5.00/acre, depending on the type of claim. In response to concerns about this patenting system, since 1994 the Congress has imposed annual moratoria on patenting via the annual Department of Interior appropriation bills.

10 A mineral is a chemical element or compound that occurs naturally in the Earth's crust. Deposits of minerals that are mined are called ores.

11 Ely, Northcutt. "Mineral Titles and Concessions." The American Institute of Mining, Metallurgical, and Petroleum Engineers, Inc. *Economics of the Mineral Industries.* Edward R. Robie, Editor. (The Maple Press, 1964.) pp . 81-130.

12 There were also various homestead acts and land grants (such as for railroads) and other acts which further encouraged westward migration and economic development of the west.

13 14 Stat. 86, Revised Statute § 2318, 30 U.S.C. § 21.

14 16 Stat. 217, Revised Statue § 2329, 30 U.S.C. 35.

15 A lode deposit is a vein of valuable mineral that resides in hardrock (or quartz); a placer deposit is an alluvial deposit of a valuable mineral.

16 Also, for some minerals the federal government earlier sold the lands outright instead of leasing them, and the question of royalties did not arise.

17 In 1994, in response to concern about the "giveaway" of federal lands, the Congress began to impose annual moratoria, via the Department of Interior Appropriation bills, on the patenting of public mineral lands.

18 Leshy, John D. *Testimony at the Hearing on HR. 2262, the Hardrock Mining and Reclamation Act of 2007.* Subcommittee on Energy and Mineral Resources, House Committee on Natural Resources. July 26, 2007.

19 For example, under the Coal Lands Act of 1873, a separate claim/patent system was created for federal coal lands. The law allowed the purchase of public coal land for ten to twenty dollars per acre, depending upon its distance from a railroad, and established guidelines for staking claims for coal mines on federal lands. See U.S. Department of the Interior. *Federal Market Value Policy for Federal Coal Leasing.* Report of the Linowes Commission. February 1984.

20 Coal royalties were not an issue prior to 1920 because with the exception of coal-bearing lands in Alaska, federal coal-bearing lands were sold rather than leased. The Alaska Coal Leasing Act provided for a minimum royalty of 2¢/ton beginning in 1914.

21 For a discussion of the history and economics of coal leasing policy see CRS Report 83-169, *A History and Economic Analysis of Federal Coal Leasing Policy*, by Duane A. Thompson and Dennis Zimmerman.

[22] According to a 1974 coal leasing study by the Council on Economic Priorities, the DOI had begun to apply ad-valorem royalties somewhat before 1976. See Council on Economic Priorities *Leased and Lost: A Study of Public and Indian Coal Leasing in the West.* (New York, New York, 1974. p. 27.

[23] The 12.5% royalty has been standard practice in the oil and gas business.

[24] There are exceptions to this general rule, for lands that are not subject to the Mining Act of 1872, such as acquired lands. In these cases, which are not frequent, miners pay advalorem rates typically ranging from 4% to 6%.

[25] CRS Report RS22567, *Royalty Relief for U.S. Deepwater Oil and Gas Leases*, by Marc Humphries.

[26] There are examples of profit sharing, instead of revenue sharing, such as in the movie business. But these reflect the reality that the return to labor (wages) could be paid in different forms.

[27] In theory, the demand schedule for factors of production by a producer of an exhaustible resource, such as hardrock minerals, is based on an amount somewhat less than the value of the marginal product, a difference that reflects the sacrifice of future profit due to present production. This causes the extractive firm to produce at an output level consistent with minimum average cost rather than the point where price equals marginal costs. Some economists believe the stock or reserves of hardrock minerals is so large relative to production or supply that the present value of sacrificed future profit due to present production is very small, so that the difference can be disregarded. Regardless of which solution actually applies, it is the market price of the mineral product that governs the demand for land and other factors of production. See Sweeney, James L. The Economics of Depletable Resources." *Review of Economic Studies.* Vol.44, No. 136. February 1977. pp. 125-141.

[28] As is discussed below, it is not always easy to define what is "at the mine," nor to determine when value is first defined and created in the case of hardrock minerals.

[29] The mining portion of the total production involves extraction and milling. Mines extract the raw mineral ore; and mills — which are usually near the mine and are considered part of the mining operation — undertake the initial processing, mainly beneficiating to produce a concentrated ore.

[30] As discussed in more detail in the last section, the percentage depletion allowance is a deduction for part of the mining firm's capital costs of doing business.

[31] The idea of using the current tax definition of gross income as the basis for the royalty on locatable minerals appears to have originated with Dr. Sandra L. Blackstone, a mineral lawyer and economist. See Blackstone, Sandra L. *Royalties for Locatable Minerals on Federal Lands: A Proposed Approach.* Statement by Dr. Sandra L. Blackstone, before the Mineral Resources Development and Production Subcommittee of the Senate Committee on Energy and Natural Resources. September 13, 1990.

[32] Some metals are produced from recycled metals.

[33] This is the average price in October 2007 on the commodities exchange. See **Table 3** on p. 29.

[34] Copper ores may also contain gold, silver, and other valuable minerals, which would affect the value of the ore concentrate per tonne (or per lb).

[35] As with the term gross income, net smelter return is somewhat misleading because it does not really refer to a return as in rate of return. In normal business parlance, rate of return refers generally to profits (i.e., gross revenue less all costs, including capital cost), while net smelter return refers to a net price, i.e., a gross mineral price less certain operating and processing costs, but not all costs. Note, however, that while there is a generally accepted standard definition of net smelter return, in fact the precise definition — and the deductions (what gets subtracted) from the gross smelter sales price — is particular to each royalty lease contract.

[36]Cartwright, Michael. R. *Mineral Production Royalties.* Available on [http://www.minval.com/royalty_mineral.html]. Note: fob denotes "free on board," which means that the price of the mineral does not include the cost of insurance or freight or other delivery charges.

[37] Barton, B.J. *Canadian Law of Mining.* Calgary: Institute of Resources Law, 1993. p. 461.

[38] Baldwin, Christopher G. *Understanding Royalty Structures.* Paper written for the Conference Board of Canada. 2003. p. 6.

[39] Anglo American Uranium Corporations SEC Filing of Form 20F12G, February 8, 2007. Available from the Securities and Exchange Commission (SEC) website.

[40] Open Energy Corporation's SEC Filing of Form SB-2/A on December 2, 2002. Available from the SEC website.

[41] To simplify the calculations, the figures are all rounded to the nearest whole number. As before, the remaining 70% of the copper ore concentrate comprise sulfur, iron, and other impurities that have no value.

[42] Prior to this law, royalties on coal were not an issue since the federal government sold, rather than leased, federal coal-bearing lands. This policy was codified by an 1864 statute, and continued as an exception to the locatable minerals policy of the General Mining Law of 1872. See Leshy, John D. *The Mining Law: A Study in Perpetual Motion.* Resources for the Future, Washington, DC. 1987. p. 30.

[43] Additional detail on the various pros and cons of various royalty bases is discussed in: Otto, Andres, Cawood, Doggett, Guj, Stermole, Stermole, and Tilton. *Mining Royalties: A Global Study of Their Impact on Investors, Government, and Civil Society.* The World Bank. 2006. Pp. 49-70.

[44] Code of Federal Regulations (C.F.R.). Title 30 Vol.2, Part 206 (Revised 7-1.2007).

[45] Ibid.

[46] U.S. Department of the Interior. Minerals Management Service. *MMS Minerals Revenue Management: General Federal and American Indian Mineral Leasing Terms*. September 2006. p. 4.

[47] Federal Register, Vol. 72. No. 84, May 2, 2007. Department of Interior. Minerals Management Service. *Geothermal Royalty Payments, Direct Use Fees, and Royalty Valuation: Final Rule*. 30CFR, Parts 202, 206, 210, 217, and 218. pp. 24448-24469.

[48] U.S. Department of Interior. *Economic Implications of a Royalty System for Hardrock Minerals*. August 16, 1993.

[49] U.S. General Accounting Office. *Mineral Royalties: Royalty in the Western States and in Major Mineral Producing Countries*. GAO/RCED-93-109. March 1993.

[50] U.S. Department of the Interior. Minerals Management Service. *Mineral Revenues 2000: Report on Receipts from Federal and American Indian Leases*. p. 134.

[51] Otto, Andres, Cawood, Doggett, Guj, Stermole, Stermole, and Tilton. *Mining Royalties: A Global Study of Their Impact on Investors, Government, and Civil Society*. The World Bank. 2006.

[52] To be technically correct, at least from an economic perspective, a lower royalty rate would be justified if the external benefits would derive from consuming the mineral, if they were proportional to price, and if they were invariant with the level of output. In some cases, re-mining a prospect with remediation of damage from earlier mining may be in the public interest.

[53] See National Research Council. *Minerals, Critical Minerals, and the U.S. Economy*. 2007.

[54] The U.S. oil stockpile is called the Strategic Petroleum Reserve. Under the authority of the Strategic and Critical Materials Stock Piling Act of 1979, as amended, the Department of Defense maintains a stockpile of about 35 strategic and critical materials to supply the military, industrial, and essential civilian needs of the United States for national defense.

[55] P.L. 95-87, Title IV. 30 U.S.C.1231 et. seq. U.S. Library of Congress. CRS Report RL32993, *The Abandoned Mineland Reclamation Fee on Coal*, by Nonna A. Noto.

[56] The largest of these is BHP Billiton, which has operations in the United States. In 2006 this multinational had a market capitalization valued at nearly $120 billion. See PricewaterhouseCoopers. *Mine, Riding the Wave: Review of Global Trends in the Mining Industry, 2007*. Available at its website, [http://www.pwc.com].

[57] U.S. Department of Commerce. Bureau of Economic Analysis. *New Release: Gross Domestic Product by Industry, 2006 (advance)*. April 24, 2007.

[58] Op. Cit. *Mine, Riding the Wave: Review of Global Trends in the Mining Industry, 2007*. Available at its website, [http://www.pwc.com].

[59] U.S. Department of Commerce. Economics and Statistics Administration. U.S. Census Bureau. *Quarterly Financial Report for Manufacturing, Mining, and Trade Corporations: 2007*. AFR/07/Q1. June 2007.

[60] The fees currently charged the hardrock mining industry are discussed later in this report.

[61] The 15% rate applies to the first $50,000 of profits, 25% to the next $25,000 of profits, and 35% to profits from $75,001 to $10,000,000. Small "non-corporate" mining companies would be taxed as sole-proprietorships.

[62] However, there are proposals to repeal this deduction as part of the Congress's tax reform proposals.

[63] This deduction was enacted under the American Jobs Creation Act of 2004 (P.L. 108-357, also known as the "JOBS" bill). It was originally a substitute for repeal of the export tax benefits under the extra-territorial income tax exclusion, which was ruled to be in violation of trade laws. See CRS Report RL32652, *The 2004 Corporate Tax and FSC/ETI Bill: The American Jobs Creation Act of 2004*, by David L. Brumbaugh.

[64] Corporations are currently taxed at 15% of the first $50,000 of taxable income, 25% of the taxable income from $50,001 to $75,000, 34% of the taxable income from $75,001 to $10 million, and 35% of taxable income above $10 million.

[65] If the statutory depletion rate is 22%, and if this were to be reduced by 20%, then the effective depletion rate would be 17.6% (22% minus 22% times 0.2). Because the statutory rate is reduced not by the full 20% but by 20% of the excess of percentage over cost depletion, the effective rate is somewhat higher than 17.6%. The specific effective depletion rate, therefore, depends upon the cost depletion amount.

[66] These calculations ignore the effects of the 50% taxable income limit on the percentage depletion allowance.

[67] U.S. Congress. Joint Committee Print. *Estimates of Federal Tax Expenditures for Fiscal Years 2007-2011*. Prepared for the House Committee on Ways and Means and the Senate Committee on Finance by the staff of the Joint Committee on Taxation. September 24, 2007.

[68] U.S. Congress. Congressional Budget Office. *Taxing Capital Income: Effective Tax Rates and Approaches to Reform*. October 2005. Table 2, pp. 10,11.

[69] Note that hardrock mining operations already have to comply with the National Environmental Policy Act, the Clean Water Act, and several other environmental laws.

INDEX

A

B

C

F

G

H

I

N

O

P

Q

R

S

U

V

W

Y

Z